C000051214

Science Based Activism

Per Espen Stoknes and Kjell A. Eliassen (Eds.)

Science Based Activism

Festschrift to Jorgen Randers

FAGBOKFORLAGET

ISBN: 978-82-450-1866-0

Graphic production: John Grieg AS, Bergen
Cover design by Fagbokforlaget
Cover photo: Tommaso Cerasuolo, artwork for the documentary Last Call by
E. Cerasuolo, prod. Zenit Arti Audiovisive/Skofteland Film, 2013, with kind
permission from Massimo Arvat

Inquiries about this text can be directed to:
Fagbokforlaget
Kanalveien 51
5068 Bergen
Tel.: 55 38 88 00
Fax: 55 38 88 01
e-mail: fagbokforlaget@fagbokforlaget.no
www.fagbokforlaget.no

Tabula Gratulatoria

Knut H. Alfsen

Dag Andersen

Harald Magnus Andreassen

Peder Anker

Iulie Aslaksen

Bent Erik Bakken

Margo and Ian Baldwin

Nicholas Banfield

Guri Bang

John Kornerup Bang

Silje Bareksten

Sven Barlinn

Anders Barstad

Jo Bech-Karlsen

Rasmus Benestad

Thor Bendik Berg

Jeanett Bergan

BI Norwegian Business School,
    Alumni Relations Office

BI-Fudan MBA program

BjR International

Hanne Bjurstrøm

Arnstein Bjørke

Torjus Folsland Bolkesjø

Svein Ole Borgen

Tore Bråthen

Hans Chr. Bugge

Steinar Bysveen

Hilde and Sven Sejersted Bødtker

Catherine Cameron

Dag Roar Christensen

Werner H. Christie

Tom Colbjørnsen

Anthony Coleman

Øystein Dahle

Pål Ingebrigt Davidsen

Jan Dietz

Caroline D. Ditlev-Simonsen

Tove Margrethe Dyblie

Knut Ebeltoft

Anders Eckhoff

Kjell A. Eliassen

John Elkington

Caroline Brun Ellefsen

Erik Engebretsen

Daniel Erasmus

Katerina Eriksen

Leif Kolbjørn Ervik

Nicolas Escalante

Stein A. Evensen

Fagbokforlaget

Nils Faarlund

Audun Farbrot

Bogdana Fedorak

Einar Flydal

Hans Kåre Flø

Kjersti Fløgstad

Christine Fløysand

Trine Folkman

Gudleiv and Trine Forr

Runar Framnes

Lasse Fridstrøm

Paul Fulton

Inge Olav Fure

Christen Furuholmen

Jostein Gaarder

Vita Galdike

Marte Gallis

Per Arild Garnåsjordet

Paul Gilding

Carl Christian Gilhuus-Moe

Gullbrand Gillund

Runhild Persensky Giving

Ole Gjems-Onstad

Ulrich Golüke

Ingunn Grande

Rob Gray

Sidsel Grimstad

Christian Grorud

Tim Greger Bådstad Hagen

Sissel Hammerstrøm

Nikolai Aurebekk Handeland

Rasmus Hansson

Bjørn Kj. Haugland

Per Ingvar Haukeland

Margit Haukås

Leif Hegna

Dag Rune Heien

Ørjan Hellstrand

Arild Hermstad

Gudmund Hernes

Christopher Hoelfeldt-Lund

Paul Hofseth

Erik Eid Hohle

Gerd Holmboe-Ottesen

Truls Holthe

Bjart Holtsmark

Øystein Hov

Bo Hu

Stig Hvoslef

Anders Gravir Imenes

Ellen A. Jacobsen

Erik W. Jakobsen

Nina Jensen

Carlos Joly

Tone Skau Jonassen

Anne Jortveit

Steffen Kallbekken

Siri M. Kalvig

Per Ivar Karstad

Franz Knecht

Peter Kramer

Knut-Andreas Kran

Idar Kreutzer

Jan Kristensen

Gunn Kristoffersen

Hilde Kristoffersen

Nigel Lake

Karen Landmark

David C. Lane

Ove Juel Lange

John Christian Langli

Mark Lee

Ulrika Leikvang

Charlotte Hartvigsen Lem

Torstein Lien

Oddvar Lind

Jonathan Loh

Kjetil Sager Longva

Ane Sager Longva og Martin Hjelle

Finn Bjørnar Lund

Johan C. Løken

Lynda Mansson

Marit Sjøvaag Marino

Ole Mathismoen

Cecilie Mauritzen

Lars Mausethagen

Desmond McNeill

Atle Midttun

John Morecroft

Ole Th. Aleksander Mortensen

Sam Mostyn

Gunther Motzke

Erling Moxnes

Naturvernforbundet Oslo og
    Akershus (NOA)

NCSC Norge

Ingun Bjerke Nikolic

Jan-Evert Nilsson

Atle Nordli

Petter Nore

Trygve K. Norman

Victor D. Norman

Harald Norvik

Tom Nysted

Inger G. Næss

Johan Fr. Odfjell

Linda Rundquist Parr

Pavlina Peneva

David W. Peterson

Joni Praded

Karen Beate Randers

Engelke Randers and Arne
    O'Donoghue

Engelke Randers sr

Alex Rau

Harald Rensvik

Torger Reve

Eliot Rich

Mats Rinaldo

Audun Rosland

Pål Rydning

Tom Vidar Rygh

Anette Rønnov

Marie Sager

Thina Margrethe Saltvedt

Yang Sang

Hans-Jürgen Schorre

Liv and Anders Seim

Harald Siem

Lukas Sihombing

Jo da Silva

Henrik Sinding-Larsen

Reidun Sirevåg

Nick Sitter

Anne-Beth Skrede

System Dynamics Society

Carl Arthur Solberg

Siri and Halvor Stenstadvold

John Sterman

Per Espen Stoknes and Anne Solgaard

Halvor Stormoen

Per Strutz

Carsten Tank-Nielsen

Anne Carine Tanum

Kristin Thorsrud Teien

Gunnar Tellnes

Einar Telnes

Anne-Christine Thestrup
Andrew Lee Thomson
Asbjørn Torvanger
Chris Tuppen
Andreas Tveteraas
Jens Ulltveit-Moe
Arild Underdal
Ingrid Elisabeth Vedeler
Arno Vigmostad
Iikka Virkkunen
Geir Vollsæter
Mathis Wackernagel

Anne Welle-Strand
Trond E. Wennberg
Fred Wenstøp
Rita Westvik
Anne Wichstrøm and Dag Gjestland
Inger Johanne Wiese
Anders Wijkman
Østfoldforskning AS
Silje Åbyholm
Carlo Aall
Lars Harald Aarø

Foreword

Festschrift for Jorgen Randers
Science Based Activism

It is a privilege to be invited to contribute the foreword to a Festschrift that honors the career and ongoing contribution of Jorgen Randers, who will shortly celebrate his 70th birthday. His dynamism and enthusiasm makes this more a mid-life milestone than a retirement for him.

Jorgen has always shared his intellectual gifts with extraordinary generosity in his teaching, research and activism. That is what makes this volume so appropriate – it is a gift to Jorgen that, by its very nature, is a gift to the wider society, containing, as it does, the work of many eminent authors on the subjects he has held close to his heart.

I have known of Jorgen's work for years, but only had the pleasure of meeting him properly in 2013, when I was immediately struck by his persuasive style and ability to bring great clarity to complex issues. He is passionate about sense and logic, as I learned at our DNV GL roundtable in Copenhagen that year which saw leading thinkers tackle the subject of "Moving Beyond Business as Usual." Jorgen was in action again at a memorable roundtable in June 2014 which we held as part of our 150th anniversary program at our DNV GL headquarters in Høvik. Headlined "The Road Less Traveled," the roundtable highlighted the critical need for a new narrative for the sustainability story – one that changes mindsets and inspires real action.

Jorgen has indeed journeyed on a less-traveled, difficult road, which has required stamina and courage over the years. I was a young scientist in the 1970s, and in common with many of my contemporaries in the wake of the anti-Vietnam war protests and the 1968 intellectual awakening, I was aware that theorizing wasn't enough.

For the eager young physicist and management graduate at MIT, co-authoring "Limits to Growth" in 1972 was a watershed. He quickly came to realize that revealing facts was not in itself enough to budge public opinion, let alone galvanize action – even facts in the form of a frightening model showing how the world's growth trajectory was on a collision course with its finite resources. Jorgen has spent a lot of his life since then urging academia, civil society and business to take action to stop the juggernaut of unsustain-ability. He has maintained this activist stance despite any limitations it may have imposed on his career as an academic and at the cost perhaps of friend-ships. While there may be no fixed definition of what science based activism is, I would like to think that Jorgen personifies it. His quest is not merely to interpret the world, but to change it.

I have always thought of myself as a scientist, and yet for much of my career I have either worked as an engineer or led them. Science is not engineering. If science is a quest to understand the universe, engineering is grounded in *terra firma* – making things work, solving problems. Engineers draw on scientific principles of course, but not always: the steam engine was well established long before the laws of thermodynamics were.

It is tempting to think of Jorgen's career as being essentially that of an engineer: a scientist, yes, but one always on a quest to improve things. Indeed, MIT's System Dynamics Group, his alma mater, was founded by Jay W Forrester – an engineer who used engineering control theory to understand complex socio-technical systems such as urban development and the world resources. I, and indeed DNV GL, feel a part of this journey with a concept we have called "taking the broader view." It means understanding and taking a stand on systemic challenges that require a broader perspective than many engineers aspire to.

For Jorgen, the time for issuing warnings about a climate crisis has passed. *2052 – A Global Forecast for the Next Forty Years* tells a sobering story of a world muddling along, focused on the short term and slowly sliding into a resource and climate maelstrom. But with Jorgen, passivity is not an option: he is still irrepressibly focused on solutions, particularly those that are attractive in the short term, but have long-term positive consequences.

My hope is that Jorgen will continue to inspire coming generations to take positive action …

Oslo, March 2015
Henrik O. Madsen
CEO DNV GL
Chairman of The Research Council of Norway

Preface &
Acknowledgements

Science Based Activism has been written by a network of researchers who have been colleagues and friends of Jorgen Randers for many years. Many are working in academic fields as well as crossing over into practical activism, in ways that he has helped shape and influence over the last four decades. The contents of the different sections and chapters are presented in the introduction.

In order to put together an edited volume, editors inevitably rely on help and assistance from a number of people. This has definitely also been the case for this volume.

Many of the authors have contributed more than their individual chapters, and we have benefited from their input in designing the volume, as well as in elaborating the themes discussed in the introduction. Sven Barlinn of the publisher *Fagbokforlaget* deserves special thanks for his resolute and professional assistance in saving the project and finding a way out of some rather fundamental difficulties. We would also like to thank Bogdana Fedorak for her assistance in the different phases of the editorial process, for overseeing communication with the contributors and together with the publisher preparing the Tabula Gratulatoria. A special thanks also goes to the language consultant, Diane Oatley.

We extend our most sincere thanks to DNV-GL with Henrik Madsen and Bjørn Kj. Haugland for co-arranging the seminar of 21 May 2015 on Green Economics at the Norwegian Business School, where this Festschrift was launched.

We would also like to thank the BI Norwegian Business School's Center for Climate Strategy, the Department of Accounting, Auditing and Law, as well as the management of the BI Norwegian Business School for supporting the Festschrift and the seminar.

Oslo, March 2015
Per Espen Stoknes and Kjell A. Eliassen

Contents

Introduction

Kjell A. Eliassen and Per Espen Stoknes

SCIENCE BASED ACTIVISM

This Festschrift has two core topics: *Science*, which in the case of Jorgen Randers is focused on system dynamics modeling of societal development, economic cycles and global sustainability. The second topic is why and how to use scientific knowledge as a basis for *activism* in the environmental, political, and business domains. Randers has been busy weaving these two strands together for almost 50 years.

Since these two topics and the relations between them are so closely interlinked in Randers' work, this volume follows closely his intellectual development and main thematic focuses over time, maybe even more than is normal in a Festschrift! In doing so, we're arguing that the best answers to the question "what is science based activism?" may be found in following in the footsteps of Jorgen himself.

What is science then in this respect? First and foremost it refers to high-level modeling as an approach and as a research tool. The advantage of high-level models is that one can make a synthesis of a large, often interdisciplinary body of research. The model then draws a map of both the past and the plausible futures that are needed for giving activism a clear direction. "Street smart" activists often rely more on broad synthesis than on in-depth empirical research. In high-level modeling you can do both. This could well be why Randers fell so completely in love with this approach already in his formative years at MIT in the 1970s.

What about activism? And what type of activism are we discussing here in relation to the work Randers has been engaged in? Activism can and has taken many forms over time and has been aimed at both political and business leaders as well as the public at large. Randers has always been creating strategies involving a broad set of stakeholders because the environmental causes need it: A substantial refocusing is required in the attitudes and actions of both politics and business, as well as an increased understanding and real action from the general public. Strategies to combat climate change are the perfect example. Here, multilevel involvements and actions have been, remain today and for the foreseeable future will be needed. This means that one cannot go too deeply into any specialist discipline. We have to remain "broad but shallow generalists" as Randers has put it. This sets up a tension between the sciences, which want to go deep and detailed on the one hand, and activism, which wants to stay practically relevant and accessible to broad groups, on the other. This volume is about exploring how to endure that tension, and the flourishing that might ensue from their combination!

From a basis in science, what type of activism becomes feasible? If the models show we're on an ethically unacceptable pathway, should one then simply try to stop the development? Or, rather than trying to break down or stop current development, it may be better to propose a positive alternative. But any such alternative is far from "value-free." It will conflict with the conventional view of scientists as being somehow neutral and descriptive. This raises the question of the ethical implications of scientific knowledge, and in particular knowledge about the future, the forecasts.

Therefore, the pathway from science, based on data, models and forecasts, to societal implications and policy advice is fraught with detours and potholes. The shift from "is" to "ought" is slippery. Being right in terms of the data and the validity of the model is no guarantee of getting it right in terms of the ensuing policy recommendations. Many times an "activist scientist" will be tempted to move beyond his/her area of expertise, overextending a feeling of clarity and certainty from her or his own research domain into areas where that knowledge is at best partial and incomplete.

Because of this, many scientists hesitate and stop short of spelling out any activist implications. There is little willingness on the part of scientists to do so, and even less reward from becoming a public figure by voicing opinions in the media. Many also fear the scorn of their colleagues and scientific critics if they should popularize their research too much. Thus sticking to one's lab and publishing scrutinized articles in discipline-specific journals, obscure to outsiders, seems to many researchers to be the most reasonable, academic career-friendly thing to do.

Yet, some persist in spelling out the longer term policy implications of their scientific knowledge. Examples include James Lovelock of the Gaia-hypothesis who strongly favors nuclear power because of his knowledge of how dangerous CO_2 emissions are in the climate system. Another example is Paul Ehrlich who issued very stark policy recommendations from his models of the "population bomb" (see Hernes, this volume). Yet another example is Rachel Carson with *Silent Spring* that contributed to a wave of environmental legislation.

Thus, if you're a scientist, getting the policy recommendations right as seen from your scientific expertise within a discipline or two is a risky venture. Nevertheless, it is a highly necessary field to venture into. Sometimes it is even ethically inevitable: What to do if your research uncovers that your fellow humans are unwittingly carrying out (self-) destructive actions on a large scale? If they are unaware of the dynamics within which they are − or are in danger of becoming − imprisoned, is there then not an ethical obligation to "cry wolf"? This dilemma lies at the core of the issue of science based activism, which the life of Jorgen Randers in many ways epitomizes.

The Cassandra situation is the classic reference for this: Cassandra of Troy had the gift of being able to see future devastating developments clearly, yet was unable to communicate these insights in a way that fellow citizens were willing to believe and act upon. She was scorned and rejected. Being able to say "I told you so" afterwards gives little comfort. Randers and the *Limits to Growth* team have suffered more than their fair share of "Don't be such a Cassandra" accusations. Is it − unfortunately for all of us − soon time for them to say: "We told you so"?

THE STRUCTURE OF THIS BOOK

This book reflects the different focuses and phases in the life of the person being honored. The various stages in the development of the personal life of Randers are also linked to the emerging of great societal and environmental issues like the increased skepticism about the future of atomic energy and weapons during the 1960s, the resource limitations and oil peaks in the 1970s, the sustainable development debates of the 1980s, the biodiversity and environmental concerns after the Rio de Janeiro meeting in the 1990s and the overriding issue of climate change from the turn of the century.

Both science and activism are closely linked to these issues over the time span from his period as a young physics student in Oslo in the 1960s to his becoming a global climate activist in the following decades. As editors, it simplifies our job that such a parallel development of history, scientific endeavors and activist activities exists!

Instead of organizing the book in chronological order, however, we've chosen a thematic structure, starting with Jorgen's focus on the activist approach, based on science, but very much activist in nature. The first section of the volume goes straight to the activist end of the science–activism continuum. John Elkington, a world authority on sustainability activism, raises the question of whether scientists are willing to go all the way to be politicized or even arrested in order to create systemic change. The next, Chapter 2, by Paul Gilding looks at how to achieve a decisive turnaround of the global society if and when the public finally demands it: "The One Degree War Plan – An Update". In Chapter 3, activist economist Iulie Aslaksen and colleagues tell the story of how the biodiversity indicators grew from a WWF-context, met with leading biologists and became an official "Nature Index for Norway" for guiding biodiversity policy. In the last chapter of this section by Mathis Wackernagel, "father" of the ecological footprint measure, looks more closely at the economic impacts of long-term ecological overshoot, in order to see how and when activism may be backed by economic interests and their understanding of risks.

Sections II and III both take on the currently most pressing issue for science based activism: strategies to combat global warming. For Randers this has been at the top of his agenda over the last 20 years – through his work in research modeling, teaching, advising governments and companies and the creation of centers and organizations.

Section II looks into social science research on climate change policy. In Chapter 5 Per Espen Stoknes reviews the recent evidence on subjective well-being to launch the idea of "A Happy Climate? New Stories for Climate Communication." Next, Caroline Ditlev-Simonsen explores whether the public in general can be seen as science based activists by conducting a Norwegian empirical study in Chapter 6 on "The Gap Between Attitude and Behavior in Environmental Protection." The following chapters shift from a social psychology approach to political science and how democracies deal with climate change. They look closely at the role and policies of the EU's Climate Policy because of the leading global role this organization has aimed for and to a certain extent succeeded in assuming. Chapter 7 by Nick Sitter focuses on the fascinating issue of "What does EU Energy Policy Mean for the Climate," showing how climate change policies to a large extent in the EU have become integrated with overall energy policy. In Chapter 8, Kjell A. Eliassen, Marit Sjøvaag-Marino and Pavlina Peneva study the more detailed inside dynamics of how the 2008 "20 20 20" package was agreed upon by the EU: "EUs Climate and Energy Policy – a Case Study of the Adoption of the Climate Change Package in 2008."

Section III reviews climate policy history and other major societal challenges explaining why it has been so difficult to act resolutely. In particular the section discusses the ambitions and shortcomings of Norway in its endeavor to try to be a role model country, something Randers has tried to promote since leading the governmental commission on Norway as a low-emission society in 2006. Marit Sjøvaag-Marino starts out in Chapter 9 by critically reviewing "The Recent History of Norwegian Climate Policy." Norway's lead climate negotiator over many years, Hanne Bjurstrøm then gives an inside view of "Norway's Role in Climate Negotiations since Kyoto." Gudmund Hernes (Chapter 11) concludes the section by discussing the shifting cycles of "Optimism and Pessimism in Climate Policy" over previous decades.

Section IV focuses on modeling, the foundation of Jorgen's activism. His formative years at MIT in the early 1970s were marked by the strong and lifelong influence of system dynamics modeling and the dangers confronting the future of the planet itself! The first chapter in this section, Chapter 12 by Victor Norman, NHH, discusses both the role of systems dynamics in environmental activism and the conflict between the worldviews of traditional economics and system dynamics. Chapter 13, by MIT's leading system dynamics professor John Sterman, subsequently looks into how to use modeling for educating the broad public by do-it-yourself types of simulations. Chapter 14 by Ulrich Golüke looks at an issue on which Randers has spent a lot of time and effort: using system dynamics to forecast shipping cycles models. This section then closes with an environmental concern very close to Jorgen's heart, the future of the forests as biofuels in relation to climate: Bjart Holtsmark's Chapter 15, "Is the Increased Use of Biofuels a Good Idea?"

With this broad summary of the topics covered, let us now take a closer look at each section from the perspective of science based activism.

FIRST SECTION: WHOLE EARTH ACTIVISM

In "No Limits to the Arresting Professor Randers" John Elkington praises Jorgen's extraordinary contributions to the field of science based activism. Given that new science often undermines old science, and old economic and political orders, it is easy to see why both some politicians and business people prefer scientists to stay in their labs, refraining from activism. He adds, "But surely thoughtful scientists – who are often much better informed on relevant issues than pretty much anyone else – should have the right to stand up and be counted if they think the rest of the world is failing to engage in new challenges." To some extent Randers has blurred the lines between science and activism. He states: "But I very much doubt that I am the only contributor here to want to see this particular 70-year-old gentleman continuing to agitate for system change as we move towards the seemingly distant 2050s, to which he has recently directed our collective attention."

Like most advocates for action on sustainability, Paul Gilding has days when he finds it hard to imagine the world waking up as comprehensively as it must do, if we are to effectively address the issues around sustainability and climate change. In the second chapter, "The One Degree War Plan – An

Update" he states that Randers was convinced the world still wasn't ready for the type of transformational action required to shift the global economy. Randers and Gilding once discussed *when* they thought real global action was likely to occur and what science could tell us about the implications of acting *decisively* at that stage. Would it be "too late"? They found that a one-degree target is still achievable and at an acceptable cost compared with the price of failure. So it seems it is possible to design a plan that would achieve the required reductions. The chapter presents and discusses the development and the implementation of such a plan. The question is not if it is needed or not, but if and when the world will commit to implementing it. When the sense of urgency gets strong enough, the war plan is ready!

Chapter 3 turns to the issue of biodiversity. Aslaksen et al. discuss "Measures of Biodiversity – Living Planet Index and Nature Index as Tools for Activism." Over the last century economic development and rapid population growth have led to increased pressure on ecosystems and biodiversity with 52% decline since 1970. What will global biodiversity look like in the future? To what extent can biodiversity measurement be a tool for policy for a planet under pressure? Ecosystem management targets need to balance the prioritized economic use of a given ecosystem at a given spatial level with society's view of what is a sustainable capacity to deliver the "bundle" of future ecosystem services.

In the fourth chapter: "What Are the Economic Implications of Overshoot, if Any?" Mathis Wackernagel defines overshoot as taking more from ecosystems than they can renew over time. He explains global overshoot as the "situation where, on aggregate, humanity's demands exceed planetary regeneration." The chapter explores questions such as: What does it mean not to have enough biocapacity? And what is the evidence? The Ecological Footprint and biocapacity accounts for nations, developed by Global Footprint Network, are designed to measure the extent of global overshoot. Given that we live on a finite planet with a limited biologically productive surface, and given that life competes for such space, the question is: How much do people demand from those surfaces (footprint), compared to how much the planet can regenerate on those surfaces (biocapacity)? The reasoning is that such scientific descriptions can enable action before overshoot becomes very destructive.

SECTION TWO: CLIMATE STRATEGY AND POLICY RESEARCH

The USA's climate policy has been in stark contrast to the EU's. Chapter 5 by Per Espen Stoknes opens by pointing to the latest quip from many US senate republicans. When asked about their view on climate, many say: "I'm not a scientist." By distancing themselves from climate science, the case for ambitious policy, they hope, is weakened. This seems to be the polar opposite of science based activism; a claim to ignorance as a basis for policy advocacy. The chapter then reviews the growth of a new scientific narrative inside of which this type of science-avoidance behavior becomes unnecessary, since the narrative reframes and aligns the long-term objectives of creating jobs, improving society and the climate at the same time. One can't win the public majority's support by arguing only for the sake of "environment," "climate" or the "planet" itself. But by linking climate to human well-being, personal happiness, jobs and radical resource-efficiency, the climate issue can be integrated with the top political and economic issues. This opens for a triple-win opportunity: well-being enhancing measures such as fewer working hours, better health from clean air and land, and the creation of less wasteful jobs, products and processes. The "happy climate" framework fits this need well: We want to move quickly forward to a society with more human well-being inside a benign climate. It's difficult to disagree with that vision.

Chapter 6 in this section by Caroline Ditlev-Simonsen discusses "The Gap Between Attitude and Behavior in Environmental Protection." She asks: Are the public "science based activists"? To what extent is it the case that improving the science based knowledge actually translates into changed behavior in market democracies? And who does the public blame for the perceived and real gap? Her study looks at the status in five areas – waste disposal, energy conservation, use of public transportation, use of water-saving showerheads, and purchase of organic foods – and presents results on the gap between attitude and behavior. A survey conducted in three larger companies, asking who the respondents believe are *to blame* for this gap, discusses the findings and establishes a foundation for proposals as to what can be done to reduce the gap. The chapter shows that people are open to intervention by the authorities to facilitate sustainable development. The author adds: "This finding may reflect individuals trying to excuse their own inaction, or it could provide an impetus for the authorities to engage in sustainability regulations."

The next two chapters move from a citizen level of analysis and up to a more institutional level: They look into the issue of *how democracies* can handle the climate issue. So far democracies have been slow in responding to the long-term climate challenges. Are market democracies at all capable of making sound, long-term decisions – or are they inherently too short-term given their typical four-year election cycles? Jorgen Randers has repeatedly expressed a skeptical stance on this question. But the EU is an interesting case since the EU has managed to make ambitious climate targets and policies binding for its members. This second section therefore concludes with two academic analyses of climate policies and in particular the policy-making processes at the EU level, before the two following chapters look at new strategies for engaging the broad public.

In the beginning the EU viewed energy policy as a national matter. It was only after the introduction of the Internal Market in the end of the 1980s with the drive for liberalization of public services that a common EU energy policy came of age. This happened at the same time as environment policy was added to the Treaty. The central question in Nick Sitter's Chapter 7 is to what extent the objectives of the EU Energy policy have turned out to be compatible with climate. Does meeting one goal make the other more obtainable, do these goals involve trade-offs, or are they incompatible – and what has this meant for the climate? EU energy policy may involve challenges for climate policy simply because it institutionalizes such a variety of member state energy policies – and some are bound to be more incompatible with the climate change policy agenda than others. The first part of the answer to the question "what does EU energy policy mean for the climate" is therefore that EU energy policy provides a stable, rule-based, long-term policy framework within which concerted action by democracies to combat climate change is possible. A second part of the answer is that the EU energy policy regime positively contributes to the EU and its member states' climate change initiatives because the Single European Market in energy has been designed and developed with a view to a *sustainable* energy policy.

The strongest part of this claim rests on the 2008 energy and climate package, and particularly the legal commitment all member states have made to the 20-20-20 goals. This is the focus in Chapter 8, by Eliassen et al.: "Are there inevitable trade-offs between policy efficiency, accountability and democratic governance, and if so, how much weakening of democracy and accountability should we tolerate in order to meet the climate challenge?" Is the political process in the EU sufficiently removed from the electorate to increase the probability of reaching the ambitious targets from 2008 and subsequently even more stringent targets? The chapter argues that the possibility of coaxing less ambitious member states into agreement through internal burden-sharing arrangements was central to the 2008 climate policy package. In addition to this "buying off" of less willing states, we also claim that the political capital invested through the individual political leadership of the presidency, particularly of German Chancellor Angela Merkel and French President Nicolas Sarkozy was decisive. The authors conclude that "democratic systems do have a potential to deliver ambitious targets for climate policies, and that investment of political capital, the benefits of which are reaped not only through immediate re-election but also through credibility and institutionalization of political goals, will contribute to increasing the probability of implementation of such goals."

The results in the studies of both national and EU climate policy show, however, that the situation of climate policies in democracies is somewhat like Odysseus binding himself to the mast to avoid derailing as the sounds of the sirens become too tempting. Perhaps Homer understood human nature and a belief in (individual) science based activism is too optimistic.

SECTION THREE: HISTORY AND ACTORS IN CLIMATE POLICY

The next section of this volume starts by analyzing the case of climate policy in Norway before turning to the global situation in the last chapter. Norway is a democracy with high environmental ambitions but rich in oil resources. Its potential to be a model country has been of central concern to Randers for more than two decades. But it also illustrates the problem of the virtually total inability of the politicians and in particular the public with respect to the need to act and act now, and act with strong force to reduce the warming of the planet.

Chapter 9, the first in this section, performs an empirical and critical assessment of what Norway has actually done on the basis of its commitment to the different treaties and the national agreement on combating climate change. Marit Sjøvaag-Marino argues that the history of climate policy in Norway from 1990 until 2014 is one of hot debates and issues, but continued high GHG emissions. This chapter concludes with three observations. Firstly, the author argues that certain fundamental principles have remained in place throughout the period under investigation. These are the "polluter pays" principle, and the prioritization of cost-efficiency. Secondly, more than two decades after the Norwegian parliament adopted the stabilization target for GHG emissions, we have greatly increased our knowledge about the potential for emission reductions from inland sources. What can and should be done is well known and documented. The third observation is, according to the author, that Norwegian climate policy is continually drawn between lofty intentions and *realpolitik*, i.e. the need for jobs and the side effects of an active fossil energy sector. The Norwegian petroleum industry is definitely among the polluters. However, the sector's importance for Norwegian wealth and job creation has made it all but impossible for politicians who want re-election to impose policies that would seriously threaten its existence, Sjøvaag-Marino concludes.

The following Chapter 10 gives an insider's account of Norway's role in the international climate negotiations. It is written by Hanne Bjurstrøm, from the Norwegian Ministry of the Environment, who has been a key person in the climate change negotiations for many years. She argues that Norway has been given a fairly influential role in the negotiation process, "relative to our size and share of the total global emissions." She emphasizes that Norway was one of the first countries in the world to adopt a CO_2 tax (1991) and to implement green tax reforms. The fact that Norway is not a formal member of the EU also leaves Norway in a unique position, as a country with an independent voice in the negotiations, she argues. Norway is, however, an oil and gas producing country and this constitutes a great challenge for Norway in terms of fulfilling its ambitions. The chapter tells the story as seen from a governmental representative perspective.

Taking a much broader outlook, Gudmund Hernes in Chapter 11, discusses the challenges of action on the basis of knowledge combined with the general public "mood." He argues that "In the 70 years since the Second World War, there have been wide swings in the public mood about the prospects for humanity" and the possibility for mankind to do something about the great challenges. The first was the threat of nuclear energy and in particular the nuclear bomb. From around 1960 there were signs of optimism because humans seemed able to build a better world by foresight, cooperation, education, technological change and freer trade. But then the dire predictions in *The Population Bomb,* "Tragedy of the Commons" and *Limits to Growth* came along to create pessimism, before the economic expansion and globalization of the 1980–1990s fired up an optimistic mood again. He asks if we will end up erring, so to speak, on the side of optimism or pessimism? The chapter gives no clear answer, but he argues that "we cannot change the way the forces of nature work – but we can change the ways in which humans interact with them." And returning to the main topic of the volume, he states that social science based activism can help us to understand the processes and actions based on this to bring forth a better world and make us more optimistic if we do something.

SECTION FOUR: SYSTEM DYNAMICS AND MODELING

In the final section the focus shifts from activism and climate policy, the current overriding issue in the work and activism of Jorgen, to the science base of modeling and system dynamics. What if the model you base your activism on turns out to be wrong?

First, in Chapter 12, Victor Norman discusses the differences between system dynamics modeling and the mainstream economics modeling approach.

Why do system dynamics modelers and neoclassical economist modelers arrive at very different types of conclusions – and thus different types of activism? This debate goes all the way back to 1972, and is masterfully summarized by Norman as seen from the economist side. It circles in on an issue of belief in capital substitutability: If you don't believe in the substitutability of natural capital, then read and re-read *Limits to Growth*. If you do find it reasonable that increases in manufactured capital can replace natural resources, then there really are no limits to economic growth, he claims.

In Chapter 13, "Learning for Ourselves: Interactive Simulations to Catalyse Science Based Environmental Activism," John Sterman analyzes the role of systems thinking and interactive simulation modeling of system dynamics not only for generating new knowledge, but also as he says "by catalysing the adoption of policies to address the pressing challenges we face" – including climate change. When doing experiments on real social systems is impossible, he shows how simulation is often the only way to discover for ourselves how complex systems work. Without whole system simulations, he argues that it becomes all too easy for policy to be driven by ideology, superstition, or unconscious bias.

Using system dynamic modeling, Jorgen Randers and Ulrich Golüke developed a model that managed to shift the focus of shipping investors. Golüke describes the process and outcomes in Chapter 14, "System Dynamics of Shipping Cycles: How Does it Create Value?" The shift was from external transport-demand forecasts to processes of decision-making internal to investors mental models. The process made it possible for investors to rapidly internalize their model output, thus improving their decisions. This model demonstrates the usefulness of appropriate models for influencing decision-making from within large, uncertain social-economic systems. By developing three innovations, the "deterministic backbone," the concept of "market sentiment" and the insight to splice the forecast onto average history, they extended normal scientific system dynamics modeling concepts to provide real value over many decades to investors. This is an example of "using science for (corporate) activism with a track record that by now has been proven."

There is increasing pressure on biological resources to replace the fossil fuels, particularly forests. In "Is Increased Use of Biofuels from Forests a Good Idea?" economist researcher Bjart Holtsmark models the effects of bioenergy from the boreal forests over time. The boreal forest stores approximately twice as much carbon as the tropical forest region. He argues that the processing and combustion of wood products is not climate neutral by calculating the dynamic consequences: If forest biofuels replace fossil fuel consumption, there might be climate gains in the very long term, but only by increasing the warming in the short-to-mid-term. So, only if the very long term is more important than the first century then biofuels from forests might still be preferred to fossil fuels. If this century is the most important, then biofuels from forests is a *bad* idea. He also questions if increased use of forest biofuels actually leads to a corresponding reduction in the use of fossil fuels.

THE BOOK AND THE PERSON:
2052 A GLOBAL FORECAST FOR THE NEXT 40 YEARS

There are deep tensions in the term "science based activism" which all boil down to an old problem in Western philosophy and science. This is the distinction between "is" and "ought." The descriptive and the normative have traditionally been seen as belonging to different domains, in particular due to David Hume's guillotine: "given knowledge of the way the universe is, in what sense can we say it ought to be different?" Hume completely cut the two apart, and ethics has long struggled to bring them together again.

But the philosophy of science has made it clear that no descriptive statements are completely void of values, norms or oughts. It lies implicit in everything from the funding of the science, the selection of methods, people and subject of study, to the selection of data, population, mode of presenting and communicating the results, as well as in the conclusions as seen flowing from the data.

So descriptive and normative aren't that far apart after all. All scientists also live a professional life in social institutions, and living a life happens according to certain norms. Thus, for many, life and science belong together, as in the case of Randers. If the "is" tells you that we're sawing away at the very branch we're sitting on, then the "oughts" become pressing to the point of it not being possible to neglect or deny them.

This volume therefore explores the concept of "Science Based Activism" from a number of angles. And all of them somehow flow out of the life and work of Jorgen Randers. With a position as an early member of the Club of Rome and the key global climate activist he has become today, one striking refection is that he, like the issue at stake, is more accepted, known and listened to at the world level than in Norway. This is why he has responded so vigorously to the need for whole earth activism in the coming decades leading up to *2052* as described in his latest book with the same title.

Part I Whole Earth Activism

The first section of this volume focuses on the activism portion of Jorgen Randers' work. In Chapter 1, John Elkington looks at science based activism as an individual choice: how far do the ethical implications of scientific knowledge require one to go? All the way to jail? Paul Gilding in Chapter 2 looks forward to the moment when the world will finally make the commitment required to avoid a full climate meltdown, and provides the ultimate activist plan in preparation for it. In Chapter 3 Iulie Aslaksen et al. demonstrate how to monitor the development of biodiversity and how indicators can be better integrated into deliberation, management and policy processes. Finally, in Chapter 4, Mathis Wackernagel explores the shift from today's demand-driven market to one of a supply-limited global market in which there is too little bio-capacity to go around as the century progresses. He addresses in detail the economic impacts of a long-term ecological overshoot, in order to see how and when activism may be backed by economic interests own understanding of risks.

These four chapters address four kinds of Whole Earth Activism, each from the perspective of the individual author: the scientist's personal activism, global-level activism, activism for biodiversity policy, and how to prepare for an era of global bio-capacity deficits. In the chapters of this section a double cry can be heard: both for more activist scientists and for a more scientific activism in the future.

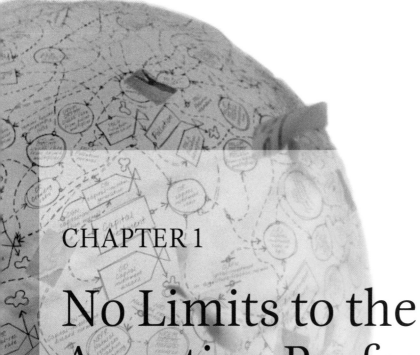

CHAPTER 1

No Limits to the Arresting Professor Randers

By John Elkington

INTRODUCTION

I will come to praise Jorgen Randers' extraordinary contributions to the field of science based activism, but to do that task justice I have decided to creep up on my target from the edge of his visual field.

And before I even begin that journey, a public health warning. I have never been a scientist – indeed, and against all the advice of my school, I gave up the study of individual sciences when I was 14, switching to General Science. At the time, I struggled to get my brain around Physics and Chemistry as they were then taught, and when it came to Biology, I refused to cut up animals. Ecology was not even on the menu.

But that didn't stop me from being interested in science and technology. And almost two decades later, I found myself writing for *New Scientist* for a number of years and for 15 years (among many other things) edited a biotechnology newsletter. In the process, I interviewed some of the world's leading scientists, including several Nobel Prize winners. Slightly to my surprise, but also to my delight, I discovered that most of these people were human, often engaging and, in a surprising number of cases, engaged (albeit generally in an understated way) in a range of ethical, social and environmental causes.

1.1   SCIENCE CAN BE A DESTABILIZING FORCE IN SOCIETY

Given that new science often undermines old science, and old economic and political orders, it is easy to see why both politicians (and the business people I have mainly worked with) prefer scientists to stay in their labs, refraining from activism. As Clive Hamilton (2010), Professor of Public Ethics, Center for Applied Philosophy and public ethics, explained with respect to the impact of climate science:

> *Innocently pursuing their research, climate scientists were unwittingly destabilising the political and social order. They could not know that the new facts they were uncovering would threaten the existence of powerful industrialists, compel governments to choose between adhering to science and remaining in power, corrode comfortable expectations about the future, expose hidden resentment of technical and cultural elites and, internationally, shatter the post-colonial growth consensus between North and South.*

Clearly, at the same time, there are good scientific reasons for encouraging scientists to stick to science, to "walk the line" of pure objectivity and to remain independent, rather than erupting into the wider world as activists – as people like Barry Commoner, Paul Ehrlich and James Lovelock have done over the years. In the latter case, however, it has been interesting to see the beginnings of a crossover between Lovelock's thinking on Gaia Theory and the wider world of business, finance and Economics (Elkington 2014).

But, at the same time, surely thoughtful scientists – who are often much better informed on relevant issues than pretty much anyone else – should have the right to stand up and be counted if they think the rest of the world is failing to engage in new challenges?

1.2   TO ACT – OR NOT TO ACT?

These seismic fault lines were illuminated in a 2013 exchange between Ian Boyd (2013a, 2013b), Chief Scientist at the UK government department DEFRA, and George Monbiot (2013), an activist and journalist. Boyd (2013a) kicked off the discussion by explaining that:

> *Any position taken by a scientist is usually a low-dimensional view of a multi-dimensional (or complex) problem … Policy-making is a messy, sometimes chaotic, process because it needs to include social, electoral, ethical, cultural, practical, legal and economic considerations in addition to scientific evidence.*

He warned that:

> *When scientists start to stray into providing views about whether decisions based upon the evidence are right or wrong they risk being politicised … If scientists start to say one or other option is right or wrong then they are beginning to take the position of politicians and they devalue the scientific evidence they claim to present. (Boyd 2013a)*

George Monbiot (2013) was having none of this:

> *Boyd's doctrine is a neat distillation of government policy in Britain, Canada and Australia. These governments have suppressed or misrepresented inconvenient findings on climate change, pollution, pesticides, fisheries and wildlife. They have shut down programmes that produce unwelcome findings and sought to muzzle scientists.*
>
> *To be reasonable, when a government is manipulating and misrepresenting scientific findings, is to dissent. To be reasonable, when it is helping to destroy human life and the natural world, is to dissent. As Julien Benda argued in "La Trahison des Clercs," democracy and civilisation depend on intellectuals resisting conformity and power.*

Boyd (2013b) pushed back, saying of scientists that:

... it is not their job to make politicians' decisions for them – when scientists start providing opinions about whether policies are right or wrong they risk becoming politicised. A politicised scientist cannot also be an independent scientist.

[Monbiot's] adversarial politicisation of science is not the solution. Instead, the scientific community needs to understand its role in the complex field of policymaking to ensure that its contribution can be maximised. This means sticking to the scientific evidence and clearly explaining the risk associated with different policy options. It does not mean passing off personal opinion as scientific evidence.

Obviously, this is a game of ping–pong that can be played almost indefinitely. But a growing number of scientists have been wrestling with their consciences as the evidence of accelerated global warming builds around the world. And it is appropriate that they do so. Science is never values–neutral, as Thomas Kuhn and others have argued over the decades. New scientific paradigms are shaped by values – and, when successful, spread them. To be aware of those values, and to critically assess them through time, is the mark of a true scientist.

1.3 ULTIMATE BADGE OF HONOR
– TO BE ARRESTING OR TO BE ARRESTED?

Two scientists who have stepped across the line into activism are Jason Cox and James Hansen. When interviewed during a 2011 protest against the Keystone XL pipeline, Cox explained (McGowan 2013):

I couldn't maintain my self-respect if I didn't go. This isn't about me, this is about the future. Just voting doesn't seem to be enough in this case. I need to be a citizen also, because this is a democracy after all, isn't it?

Arrested several times, Hansen took a microphone before he was led away from a Keystone XL protest outside the White House in 2011. He implored President Obama to act "*... for the sake of your children and grandchildren*" (Drajem 2011).

With 17 fellow scientists, all using their scientific credentials in political engagement, Hansen also wrote a letter to President Obama on the Keystone pipeline, in 2013, explaining:

Dear Mr. President,
We hope, as scientists, that you will demonstrate the seriousness of your climate convictions by refusing to permit Keystone XL; to do otherwise would be to undermine your legacy.[1]

---

1   18 top climate scientists call on President Obama to reject Keystone XL, 15 January 2013, 350.org, http://350.org/18-top-climate-scientists-call-president-obama-reject-keystone-xl/

And then there was the case of Gus Speth, described at the time as the "ultimate insider," and as "a 72-year-old gentleman who has spent his adult life as an adviser to US presidents, executive of UN agencies, scholar-professor-dean at some of the world's leading universities." Having worked with Speth, I was not hugely surprised that he had been prepared to be arrested, but I was intrigued to read his account of the adventure:

> [The police] decided to use the first group of us to set an example to discourage the others. It didn't work, but the result was that they treated us pretty much like common criminals. … We ended up in a central cellblock in the D.C. jail for three days. We spent a lot of time in leg irons. Slept on stainless steel slabs without any bedding or cover or pillow or anything – just stainless steel.
>
> Ate baloney sandwiches – two a day – and water. We were fingerprinted, mug shots – I guess I have a record now. In the end they didn't press any charges against us. They just opened the door and let us walk out after three days.
>
> In fact, we had a high-spirited three days in the D.C. jail. There were 60-some-odd people there with us in jail, and they knew I was a professor, and they wanted me to give a lecture. So I gave a long lecture on the need for systemic change while there in the central cellblock in the D.C. jail. (Cohn 2013)

So have such people gone too far – or would future generations question whether they have yet gone far enough? One interesting perspective here comes from Jeremy Grantham, co-founder of investment house GMO. A long-standing advocate of more urgent and effective action on climate change, and a man who has put his (substantial) money where his mouth is, he has insisted that:

> It is crucial that scientists take more career risks and sound a more realistic, more desperate, note on the global-warming problem. Younger scientists are obsessed by thoughts of tenure, so it is probably up to older, senior and retired scientists to do the heavy lifting. Be arrested if necessary. This is not only the crisis of your lives – it is also the crisis of our species' existence. I implore you to be brave. (Grantham 2012)

As the struggle intensifies, between the old economic order and the new, between the old drill-baby-drill paradigm and the emerging, only-one-Earth paradigm, the dilemmas faced by scientists can only intensify.

The new science dictates very different priorities, directions and decisions than the old science. And, as Professor Hamilton (2010) noted, the political, economic, financial and social implications of the coming transformations are profound. It is not just coal mines that will become "stranded assets." Entire cities, regional economies and economic models will be stranded as the shift builds momentum. Whether they are arresting in their communications or communicating to the world following their arrest, scientists have a moral responsibility to speak out and, where necessary, to take a stand on the issues.

1.4 TIME TO CROWD-FUND THE JRCP?

And, to bring all of this back to the subject of this well-deserved *Festschrift*, Jorgen Randers, a man I have been privileged to encounter in many places and seasons, I have just this to say. Here is a truly eminent, thoughtful man. A man at all times well ahead of his times. A wearer of many impressive hats over the years. A scientist and an activist. And a leader who has gracefully, apparently effortlessly and very effectively inspired generations of people, among them tomorrow's scientists and activists. In the process, he has blurred the lines between science and activism.

An arresting man, in short, if not (to my knowledge) yet arrested. Still, I am confident that Jorgen's name and influence will reverberate through the decades to come, as it has since *The Limits to Growth* first appeared over four decades ago.

It is in the very nature of a *Festschrift* that it signals a passing of the baton, of the torch, to those compiling the work – and to their rising generation. That is as it should be. But I very much doubt that I am the only contributor here to want to see this particular 70-year-old gentleman continuing to agitate for system change as we move towards the seemingly distant 2050s, to which he has recently directed our collective attention (Randers 2012).

Meanwhile, just a thought, is anyone prepared to help crowd-fund the Jorgen Randers Cloning Project?

BIBLIOGRAPHY

350.org (2013). 18 top climate scientists call on President Obama to reject Keystone XL, *350.org,* http://350.org/18-top-climate-scientists-call-president-obama-reject-keystone-xl/

Boyd, I. (2013a). Point of view: Making science count in government, *eLife.* (online) DOI: 10.7554/eLife.01061

Boyd, I. (2013b). Scientists do speak up, but politicians decide policy, *The Guardian*, http://www.theguardian.com/commentisfree/2013/oct/06/scientists-speak-up-politicians-decide-policy

Cohn, R. (2013). Charting a New Course for the US and the Environment, interview with Gus Speth, *Yale E360*, http://e360.yale.edu/feature/interview_gus_speth_charting_new_course_for_us_and_environment/2612/

Drajem, M. (2011). NASA's Hansen arrested outside White House at pipeline protest, *Bloomberg News,* http://www.bloomberg.com/news/2011-08-29/nasa-s-hansen-arrested-outside-white-house-at-pipeline-protest.html

Elkington, J. (2014). Gaia Theory & Economics, *Volans,* http://volans.com/2014/04/gaia-theory-economics/

Grantham, J. (2012). Be persuasive. Be brave. Be arrested (if necessary), *Nature,* http://www.nature.com/news/be-persuasive-be-brave-be-arrested-if-necessary-1.11796

Hamilton, C. (2010). *Why We Resist the Truth About Climate Change*, paper to the Climate Controversies: Science and politics conference, Museum of Natural Sciences.

McGowan, E. (2013). Climate scientist willing to face arrest at tar sands pipeline protest, *The Guardian*, http://www.theguardian.com/environment/2011/aug/18/climate-scientist-tar-sands-pipeline-protest

Monbiot, G. (2013). For scientists in a democracy, to dissent is to be reasonable. *The Guardian*. http://www.theguardian.com/commentisfree/2013/sep/30/scientists-democracy-dissent-reasonable-boyd

Randers, J. (2012). *2052: A Global Forecast for the Next 40 Years.* White River Junction, Vt.: Chelsea Green Publishing.

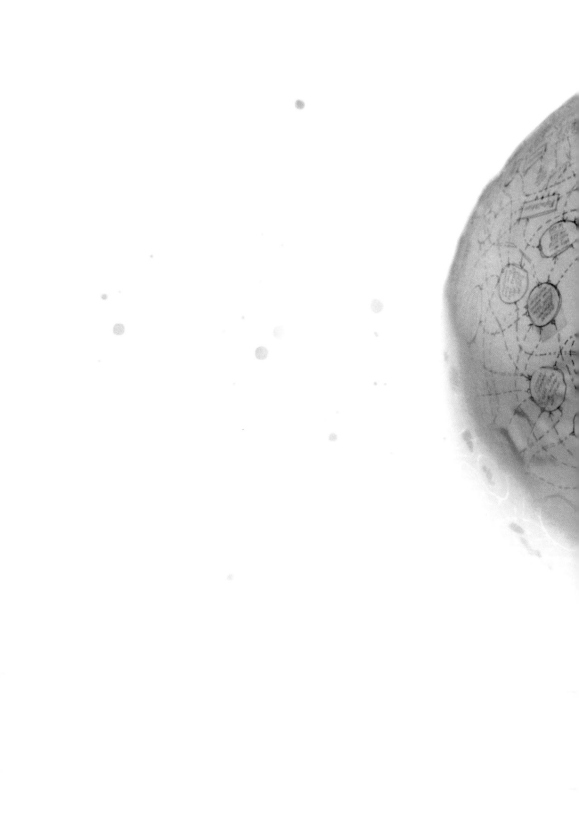

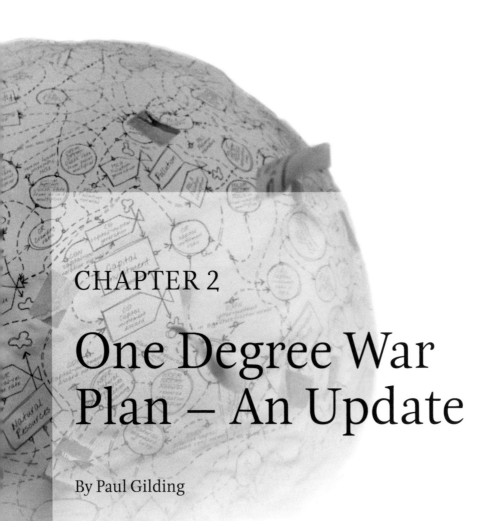

CHAPTER 2

One Degree War Plan – An Update

By Paul Gilding

INTRODUCTION

Like most advocates for action on sustainability, I have days when I find it hard to imagine the world waking up as comprehensively as it must do, if we are to effectively address the many global challenges around sustainability and climate change described throughout this book.

Yet, when I sit back and reflect on the topic, I always conclude the world will respond and that when it does, it won't be "too late." I didn't always feel that way. Like most advocates for action in this area, I shared the assumption that society was capable of letting the situation reach a point where it would be "too late," a point where we would not be capable of stopping a runaway process of ecological collapse and with it the collapse of civilization as we know it.

This risk of a runaway breakdown is perhaps the most important issue in this whole area. It is of great concern to scientific experts seeking to understand whether there are tipping points where the global ecosystem takes over and acts on such a scale that nothing we do can then have any influence.

I am confident such points exist, but I am also convinced we will act before we reach them. Of course one can never know for certain given the complexities of the global ecological system but on balance that's where I land. I wasn't always so sure. It was only when I understood the speed and scale of what a true crisis response could achieve that I realized just how dramatically we could, and I believe will, respond when we do. It is not a pretty picture – and it certainly will not be a smooth ride – but I believe it is a realistic view of how this will unfold.

The turning point for me was research undertaken with my friend and colleague Professor Jorgen Randers, professor of climate strategy at the BI Norwegian Business School. As one of the original authors of the Club of Rome report *The Limits to Growth,* Randers has been a tireless advocate for action on sustainability since that book was published in 1972 and became the bestselling environmental book of all time (Meadows et al. 1972). He is deeply experienced in these issues and from many points of view. Along with his MIT PhD and his current professorial role, he has been a company director, a business school president, deputy head of the World Wildlife Fund, and an investment manager.

2.1 THE INEVITABILITY OF THE WORLD WAKING UP

Randers and I met as core faculty on the Prince of Wales's Business and Sustainability Program, an in-depth seminar for corporate executives run by Cambridge University's Institute for Sustainability Leadership, where I am a

Fellow. After one of these seminars in 2007, Randers and I, joined by my wife, took some time out and went mountain bike riding in the Barrington Tops National Park in the Hunter Valley north of Sydney. Over dinner one evening, the three of us were discussing how we saw the global response unfolding now that the economy was moving beyond the limits to growth. We had first discussed this issue a year earlier with the team at my then advisory business, Ecos Corporation, brainstorming what a global crisis or emergency response might look like.

With thirty-five years of focus on these topics, Jorgen had a great deal of wisdom to share. Indeed, in 2004 he had published, with his colleagues from 1972, the thirty-year update to *The Limits to Growth* titled *Limits to Growth: The 30 Year Update*, where they explored this very question (Meadows et al. 2004).

Around the time of our discussion, there had been greatly renewed public attention on climate change and sustainability. Governments and the corporate sector were deeply engaged in these issues, and the public, driven by major climatic events and high-profile campaigners like Al Gore and Tim Flannery, had put the issue at the forefront of public and political debate. Many experts argued we'd turned the corner and would now start to see serious political action.

Randers was skeptical of that view. He had seen the issue ebb and flow over many decades, from the 1970s oil shock through various peaks of attention in the 1980s and 1990s to the then emerging global financial crisis. He was convinced the world still wasn't ready for the type of transformational action required to shift the global economy. He mounted a convincing argument, so our conversation moved to when we thought real action was likely to occur and what the science told us about the implications of acting at that stage. Would it then be "too late"? If not, what type of response would be necessary to prevent societal collapse?

Our first conclusion was that the world was probably still a decade or so away from really engaging with a comprehensive response. We knew what this meant, given the lags in the global ecosystem and what the latest scientific research was saying about accelerating impacts. Any response that hoped, at that late stage, to stabilize the global ecosystem would have to be breathtaking in scale, certainly compared with any proposal on the table in 2007. Otherwise it would indeed be "too late" because the lagging impacts would overcome anything less. So we knew immediately we were talking about an economic and social mobilization comparable to that seen in a world war.

Two things occurred to us as we explored this idea further over the coming days, while cycling and walking through the mountains. First, there would have to be acknowledgement that this was a major global existential and economic crisis before such a response would be implemented, because nothing else would drive the dramatic shift in the political context that would be necessary. Second, we knew of no mainstream global research underway to define the response that would then be needed to be effective. All the work being done was based on what Churchill called "doing our best" rather than "what was necessary." The science was clear on what was necessary, and we knew even the most dramatic proposals on the table in 2007 didn't come close.

These conclusions gave us some important insights into how the future could unfold and also set us a clear task to take on. The fact that a crisis would be needed before society adequately responded meant such a crisis was inevitable. That's because the momentum of increasing impacts in the global ecological system would accelerate until denial was impossible, in turn triggering a society-wide crisis response. This meant the scale of response we foresaw, impossible to imagine in 2007, was not just possible but actually likely. History suggested as well that when it emerged most people would be caught by surprise.

2.2 HUMANITY IS GOOD IN A CRISIS

That in turn meant the world at that time would urgently need a well-considered crisis response plan but wouldn't have one. So we decided to start the process by writing our version of such a plan and putting it into the public domain. Over the following two years, we did so.

Our prime objective was to encourage other experts to engage on the approach, ideally motivating government policy makers to dedicate adequate resources to a comprehensive version of such a plan, even if just as a contingency. Our other objectives were to alert climate advocates, businesspeople, and the community in general that such a warlike mobilization was at least likely, and therefore we all needed to prepare for it. It was also to alert them that their perception of "too late" might need recalibrating, given what a crisis response could achieve.

We concluded our work and put it into circulation in 2010 via the academic publication *The Journal of Global Responsibility*. That paper (Randers and Gilding 2010) provides the foundation for what I present here.[1]

Back in 2007 when we agreed to write the One Degree War Plan we had to imagine a different world than the one we saw at the time. We had to

---

1   The paper is freely available on my website http://paulgilding.com/2009/11/06/cc20091106-odw-launch/

imagine a world that, having woken up to the risk of collapse, would accept that action could no longer be delayed because key tipping points could be passed that would put survival at risk.

There would have to be sufficient impacts to eliminate any serious political debate about the causes or the risks. In fact there would need to be powerful political forces, in business, the military, and the community more broadly, demanding urgent and dramatic action. This demand would have to be sufficient to overcome vested interests' fight for protection of their economic wealth.

The key shift from 2007 would therefore have to be that a critical mass *within the elites* in the economy and society more broadly had come to recognize this issue as a threat to their own power and wealth. In other words, "the 1%" had to engage.

Systems fight hard to protect themselves and some years later we see this emerging with powerful forces in the elites lining up on opposite sides. On the one hand we have the fossil fuel sector fighting for its life against the momentum for action on climate change, including divestment campaigns, the collapsing value in coal assets and oil prices, an incredible boom in solar energy and disruptive change in the electricity and auto sectors. Perhaps most critically we have on the other side aligned against them both disruptive companies like Tesla and old companies like Unilever who are arguing for strong action on climate on the grounds of economic opportunity and global economic stability. This will inevitably accelerate in intensity, as the threat to fossil fuels gets stronger, matched by the growing recognition that fixing climate change is an economic opportunity of mind-boggling proportions.

While we see this emerging, as I write this it is not *yet* of sufficient strength to overcome the resistance to change but it inevitably will be in the coming years, and I believe in this decade.

Given this, there are important lessons from the process where World War II was declared both in the United Kingdom and in the United States. There were also lessons to be learned about leadership from Winston Churchill's approach to the extraordinary challenge faced by Europe (Juniper 2014).[2]

When the question becomes survival of the economic system and global geopolitical stability, the first question to be asked will be Churchill's "what is necessary." While it is obvious that the challenge we face is much broader than climate change and goes to the essence of our socio-economic model

---

2   See a comparison of Churchill's leadership vs. what's needed today on the climate crisis here: http://www.yesmagazine.org/issues/how-to-eat-like-our-lives-depend-on-it/winston-churchill-leading-fight-climate-change

and the very idea of infinite growth, climate change will be the initial focus. There are two reasons for this. First, the system will correctly judge that climate change is the most immediate, clear, and present danger. If it is not effectively addressed, economic and social collapse will prevent anything else from being dealt with. Second, the system will incorrectly believe that we can continue with our present economic model if we decouple growth from CO_2 emissions and make our economy exceedingly more efficient in material consumption.

Given that the first point is true, however, there is great benefit in having society focus sharply on greenhouse gases and climate change. It will after all, as we will see, require an extraordinary level of focus and effort to be effective.

2.3 WHAT WOULD BE *NECESSARY* TO FIX CLIMATE CHANGE?

So with this in mind, what will be *necessary*? What would it take to "fix climate change"?

To the objective observer, the climate science is clear on what is necessary and this hasn't changed since Randers and I did our work together in 2009. The framework for this science generally translates into how many degrees centigrade we can allow the average global annual temperature to rise above the level it was before the Industrial Revolution.

This then translates into a maximum allowable level of greenhouse gas concentrations in the atmosphere to keep below that given temperature target, with the range of uncertainty reasonably well understood.

This concentration level is generally measured as CO_2e (all the main greenhouse gases converted into their equivalents in impact to CO_2, the key greenhouse gas of concern). While this science is imprecise because of uncertain feedbacks, it is currently assumed by policy makers that to have a "reasonable chance" (generally considered at around 50%) of achieving two degrees of warming, the greenhouse gas concentration in the atmosphere must be kept to less than 450 ppm CO_2e.

Although broadly accepted as the agreed goal by policy makers, including the 2009 Copenhagen Conference and hundreds of global corporations, few mainstream science groups actually argue that this is a "safe" level. Rather, it is assumed to be "the best we can do" based on the analysis of what is politically "realistic." Two degrees will in fact lead to widespread environmental, social, and economic disruption, including widespread threats to food supplies, dramatic increases in extreme weather, and a significant rise in sea level. Most importantly, we would face an unacceptable level of risk of runaway warming threatening the stability of civilization.

So two degrees of warming is an inadequate goal and a plan for failure. It is not rational, it is not based in science and it is not the type of risk management approach we take to any other human endeavor.

The logical, science based response is to set a target that gives society a "safe" outcome. Based on currently available science, bringing global warming back to below one degree centigrade above preindustrial levels can be considered reasonably "safe" for humanity on a crowded planet. It will still have negative impacts, as we are seeing today, but they will be manageable, at least based on current knowledge.

Returning below one degree of warming, in other words, is the *solution to the problem*. It is "what is necessary." Thus Randers and I concluded that when the crisis hits and the scale of the threat was understood by a sufficient critical mass of the elites and broader society, there would be demand to achieve no more than one degree of long-term warming.

2.4 IMAGINING THE IMPOSSIBLE

Given society is struggling to agree and act upon a 2 degree target, with many arguing it's impossible to achieve even that, a brief comment is warranted on the idea of going to the much tougher 1 degree target.

I find rather strange the conversation with people who argue that a target of 1 degree is "impossible" or "unrealistic." As if standing by and allowing civilization to slide into collapse and the descent into chaos, is somehow more "realistic"!

I soon find myself drawing on my favorite quote from Nelson Mandela – "It always seems impossible until it's done."

There are many things that have been achieved that seemed impossible beforehand. Certainly WWII from the point of view of the UK would easily qualify in that regard. So the question has to go back to the ideas of Churchill again – we must ask "what is necessary." And the answer to that is clear.

It was interesting that in our research we concluded that the CO_2e concentration required to achieve this one degree of warming was around 350ppm. This is the same level being called for by scientists such as James Hansen of NASA and also endorsed as the likely end target by many others. It is also the focus of many in the global climate movement, particularly around Bill McKibben's 350.org. Many scientists in the heavily politicized arena of climate understandably prefer not to enter the public debate on what is a "safe" target, given that even two degrees creates such resistance. However, I've now had enough private conversations with world-class scientists to be confident that the scientific community will before long settle on this as the upper end of the right target range.

It is interesting to consider the context of risk here. The nature of emissions reduction curves (how an end target translates into annual reductions to get there) means it's very hard to strengthen targets later. So the logical approach to uncertainty, given what's at stake, is to have a stricter target and then relax it later if the science firms up and suggests it's safe to do so. So from every rational view, one degree is the right place to start, given failure is catastrophic and probably irreversible.

I understand many still respond to such a target as "unachievable" or "unrealistic," believing we are inevitably on our way to two degrees or more. In considering this view, it is critical to differentiate between what people believe is politically "realistic," which is a subjective judgment based on the present political context, from what would be possible if we decided to address the issue with our full capacity.

In a 2010 issue of *Nature Geosciences,* two Canadian scientists used existing models to demonstrate that if we stopped all emissions tomorrow, temperatures would stop increasing almost immediately and decrease over time (Matthews and Weaver 2010). In summary, the only warming that is truly "locked in" is the warming we choose to create by continuing to emit. A separate study in *Science* (Davis, Caldeira and Matthews 2010) found that if all existing energy and transport infrastructure was used for its natural lifetime, but no new infrastructure emitting greenhouse gasses was created, warming would peak at 1.3 degrees and then start declining. Again, the conclusion is that we can physically do this – we just have to want to do it badly enough.

So if one degree is what is *necessary* and more than this is defined as the "enemy" for our "One Degree War," what action is required to win the war, and would the required action be possible to achieve? In other words:

1. Is an agreement to achieve such a plan politically conceivable?
2. If it were, is it technically and economically possible to reduce emissions to a level that will bring warming back below one degree?

Clearly, agreement to the One Degree War Plan is hard to imagine today, let alone back in 2007 when we started focusing on it. However, in both World War II and the recent financial crisis, there are clear examples of how fast things can change and how apparently intractable opposition and resistance can quickly evaporate. In the case of World War II, the speed of response by the United States was extraordinary. For example, whereas in 1940 US defense spending was just 1.6 percent of the economy (measured as GDP), within three years it had increased to 32 percent, and by 1945 it was 37 percent. Given that GDP increased itself by 75 percent in that time, the observed increases are even more extraordinary. Similarly extraordinary political decisions were made to direct the economy. For example, just four days after the bombing of Pearl Harbor, the auto industry was ordered to cease production of civilian vehicles (despite earlier commitments to ensure that this would not happen and intense earlier resistance from the companies involved).

Gasoline and tires were rationed, campaigns were run to reduce meat consumption, and public recycling drives were held to obtain metals for the war effort. Yes, there was still plenty of resistance, but the political leadership of the day, with public and business support, simply overrode it for the greater public good – because the consequence of failure was unacceptable.

So it *can* be done. But *how* would it be done? It is unlikely that the One Degree War would result from a universal global agreement. The process around the Kyoto Protocol and the Copenhagen meeting shows how difficult global agreements are. This difficulty in reaching consensus is often put to me as evidence that we will fail to act on climate change. My response is to ask, "Can you think of other examples where a major military action or economic transformation was driven by a global consensus agreement?" On what basis did we ever believe such an approach would be possible with climate change, especially when many participants have actively sought to undermine it?

We didn't seek a single global agreement to free trade before any action was taken, for example. If we had done so, we would probably still be negotiating on the preamble fifty years later! Instead we started with consultative bodies like the General Agreement on Tariffs and Trade (GATT); we negotiated agreements between individual countries and then expanded them to regions. Meanwhile, very, very slowly, we built the global infrastructure for governance of trade, starting in 1947 with the formation of GATT and continuing up to 1995 to form a body with enforcement power, the World Trade Organization (WTO). More than sixty years after GATT, even the WTO is still not global in impact, with even China joining only in 2001 – that alone took fifteen years of negotiations.

So on climate change, an even more complex economic issue and with significant business opposition to change, it is hard to imagine we would jump straight to a single, legally enforceable, global agreement even in a crisis.

2.5 THE COALITION OF THE COOLING

When we do decide to launch a rapid response, it is far more likely that a small number of powerful countries, a kind of "Coalition of the Cooling," will decide to act and then others will follow. Some will follow in order to align with the major powers, and some will do so under military, economic, and diplomatic pressure to join.

In a technical sense, this process is easy. A full 50 percent of global greenhouse gas emissions will be covered if three "countries" (China, the United States, and the EU-27) agree to act. If we add another four countries (Russia, India, Japan, and Brazil), the coalition will control 67 percent of global emissions. Add a few friends and we soon move to more than sufficient impact to tackle the problem. We saw this start to emerge in Copenhagen, and then more recently with the very significant agreement between China and the US. While such developments are always messy and will ebb and flow over coming years, there is no doubt in my mind that this is the primary way progress will emerge.

The answer to the first question is therefore clearly yes. When we accept the crisis, we are capable of taking the political decisions required to get to work on the action plan. So is there an action plan that would work?

What our work showed is that based on current knowledge and technology, *a one-degree target is completely achievable and at an acceptable cost compared with the price of failure.* It would be very disruptive to parts of the economy and to many people, and it would require considerable short-term sacrifice, but it certainly "solves the problem" and it is clearly economically viable, possibly even positive – as we'll return to.

So from both questions, our political decision-making capacity and our technical/economic capacity, the issue is not humanity's *capacity* to act, but the conditions being such that humanity *decides* to act. Identifying this point is simple: When the dominant view becomes that climate change threatens the viability of civilization and the collapse of the global economy, a crisis response will rapidly follow. Then society's framework will change from "what is politically possible" to Churchill's "what is necessary." Until then, little of real substance will happen except getting ready for that moment by building companies and technologies so that we're ready to scale up quickly – such as we have seen with solar energy.

What would such a "war plan" look like? Can we forecast the likely response that will be implemented when the moment comes? Randers and I thought so. In designing our draft plan, we estimated a start date of around 2018, not as a precise prediction, but we needed a start date to model our response and its impact, and 2018 was our best judgment on when this would emerge. Some years later, with such powerful momentum building for action in some parts of the business community and such disruptive change happening in the energy sector, I would argue this still seems like a reasonable forecast – though again not a precise prediction.

We concluded that at that late stage, four types of actions would be required to take control of the crisis:

1. A massive industrial and economic shift that would see the elimination of net CO_2e emissions from the economy within twenty years, with a 50 percent reduction in the first five years.
2. Low-risk and reversible geo-engineering actions to directly slow temperature increase, to safely overcome the lag between emissions reduction and temperature impact.
3. The ongoing removal of around 6 gigatons of CO_2 from the atmosphere per year for around one hundred years and the long-term storage of this CO_2 in underground basins, in soils and in biomass.
4. Adaptation measures to reduce hardship, inequity and geopolitical instability caused by the unavoidable physical changes to the climate, including food shortages, forced migration, and military conflict over resources.

It is a symptom of the magnitude of the task that even with the dramatic action proposed in our One Degree War Plan, warming would continue above one degree until the middle of this century, before falling back to plus one degree by 2100.

2.6 LAUNCHING THE ONE DEGREE WAR

We suggested fighting the One Degree War in three phases:

1. *Climate War. Years 1–5.* Modeled on the action following the entry of the United States into World War II, this would be the launch of a world war level of mobilization to achieve a global reduction of 50 percent in greenhouse gas emissions within five years. This crisis response would shock the system into change and get half the job done.
2. *Climate Neutrality. Years 5–20.* This would be a fifteen-year long push to lock in the 50 percent emergency reductions and move the world to net zero climate emissions by year 20 (that is, in 2038 if we start in 2018). This will be a major global undertaking, requiring full utilization of all technological opportunities, supported by behavioral and cultural change.
3. *Climate Recovery. Years 20–100.* This would be the long-haul effort toward global climate control – the effort to create a stable global climate and a sustainable global economy. Achieving this will require a long period of negative emissions (i.e. removing CO_2 from the atmosphere) to move the climate back toward the preindustrial "normal". For instance, some refreezing of the Arctic ice cap will require removing CO_2 from the atmosphere through geo-engineering actions, like burning plantation wood in power stations and storing the emissions underground using car-

bon capture and storage (CCS). We believe humanity can complete the stabilization job in the first decades after 2100.

We tested our suggested emission cuts in the C-ROADS global climate model developed by Climate Interactive, an initiative of Ventana Systems, Sustainability Institute and MIT's Sloan School of Management (See Sterman, this volume). This confirmed that implementation would deliver broadly the following results:

▶ The CO_2e concentration falls below 350ppm by the end of the century, after peaking at around 440 ppm.
▶ Global temperature does temporarily rise above plus one degree centigrade in mid-century, then falls below plus one degree centigrade around the end of this century.
▶ Average sea level rises by 0.5 meters around 2100 and continues rising to a peak of 1.25 meters around 2300. This is still very disruptive and might trigger a tighter target, but 1.25 meters over three hundred years is at least more manageable, with good preparation, than current forecasts given the longer time frames.

In broad terms, what this all means is that the climate would be stabilized and manageable for global society. There would still be substantial changes to the climate, disruption to the economy and food supplies, and great loss of biodiversity. However, it would be manageable and it would reduce the risk of the collapse to a tolerable level. It would also allow stronger action if the science indicates the situation was worse than expected.

So it seems it is possible to design a plan that would achieve the required reductions. Of course this is just indicative. What is needed is a multiyear detailed modeling and planning exercise on a scale normally only undertaken by governments. Our point was simply to show what is possible.

So what types of real-world actions does our plan indicate would be required?

We proposed a dramatic and forceful start of the One Degree War, for two reasons:

1. There is disproportionate value in early actions – because the impact of emissions is cumulative, cuts taken earlier in a program save much larger and more disruptive reductions later.
2. History indicates that successful responses to crises tend to involve urgent, dramatic actions rather than slower, steady ones. This engages the public and breaks the tyranny of tradition.

The One Degree War plan therefore proposes a series of global measures to achieve a rapid halving of CO_2 emissions during the initial five-year 1C-war, through linear reductions of 10 percent per year. The C-ROADS model indicated that it takes cuts of 50 percent in the period 2018–2023 to reach our 1C goal. Even then, this cut must be followed by reductions to zero net emissions by 2038 and net absorption, each year for the rest of the century, of 6 $GtCO_2e$/year (gigatons of CO_2 equivalents per year). While the initial 50 percent in five years is very challenging, it is certainly doable. Critically, a slower start would risk a larger overshoot and therefore make it much more challenging to achieve the one-degree goal.

The good news is that cutting by 50 percent by 2023 can be achieved with the types of initiatives that studies like those by international management consultancy McKinsey & Co indicate will cost society less than 60/tCO_2e. (ton of CO_2e). The bad news is that making these cuts at a faster speed will, by conventional wisdom, increase the cost. This is based on infrastructure having to be scrapped before the end of its useful life and because technologies will have to be implemented before they are commercially mature. If this is accurate, it is the unfortunate consequence of acting late, as we will be. Delaying action would, however, just make that worse.

There is a counterargument that was not possible for us to model, but we were inclined to support, that a warlike mobilization of the global economy to transform our energy and transport infrastructure will not only be affordable, but may in fact trigger so much innovation and economic activity that it ends up being positive economically. This is argued by many analysts in this area, who see renewable energies as so immature that they will inevitably become not just cheaper than today, but cheaper than fossil fuels even without a carbon price.

Since we wrote the plan, developments in solar energy, particularly its disruptive impact through distributed household systems, tends to support this view. We have seen an extraordinary acceleration in renewables, particularly solar, which reinforces my view that we were right on the innovation point. We've seen solar drop in cost by 80% since we started writing our plan, while electric cars are now seen as not just a novelty but many argue are the future of personal transport. And this is not just a technology view but a firm market one, with the electric car company Tesla valued at around half the market cap of GM, despite production volumes in 2013 of 30 000 units vs. GM's 9 000 000 units! There are now many situations where solar is cheaper than coal and in some cases even gas. Given that the solar revolution has just begun we can expect such developments to continue.

Certainly the types of approaches proposed in the One Degree War Plan would unleash massive innovation on a scale far larger than we've seen so far, which means this argument would rapidly be proven either way. It is the case in previous wars that innovation drove new industries and great efficiencies because the determination to achieve an outcome forced major breakthroughs in technology and overcame normal commercial development impediments.

This debate is largely of academic interest only, as the One Degree War Plan assumes a level of crisis that will dictate the approach that has to occur, largely regardless of the cost. I don't imagine there was much of a cost-benefit analysis done on the Manhattan Project when the US government decided it needed to produce an atomic bomb. So we can safely leave to history the judgment of the relative costs of CO_2 reduction.

2.7   EXAMPLES OF THE MAIN ACTIONS REQUIRED

To provide a flavor of what we can expect to see when the type of response we outline occurs, I will present some summary points from the plan. I will after each one provide some personal reflections on more recent developments by way of update. These excerpts indicate the types of actions that would be required in the first five-year period to get the global economy on the path required to ultimately bring global temperature increase below plus one degree centigrade.

CUT DEFORESTATION AND OTHER LOGGING BY 50 PERCENT

Reduce by one-half the ongoing net forest removal and land clearing across the world, including tropical deforestation. At the same time, concentrate commercial forestry operations into plantations managed to maximize carbon uptake. This will require significant payments to developing countries, for the climate services provided by their intact forests, but is surprisingly cost-effective and doable.

> *Update: While the issue has not advanced materially, it is notable that there is very strong advocacy for action in this area, including from the corporate sector which is increasingly seeing the economic risks. As I write this, Brazil's richest city Sao Paulo, with 20 million people, has just a few months of water supply left, with deforestation a major cause, made worse by climate shifts. If Sao Paulo ran out of water you can be certain the economic argument for action would accelerate rapidly.*

CLOSE ONE THOUSAND DIRTY COAL POWER PLANTS WITHIN FIVE YEARS

Close down a sufficient number of the world's dirtiest coal-fired power plants to cut the greenhouse gas emissions from power production by one-third. We estimate this implies closing down one thousand plants, resulting in a parallel reduction in power production of one-sixth. (Power production would fall proportionally less than emissions, because the dirtiest plants emit more CO_2 per unit of energy.)

Update: We've seen dramatic falls in the popularity of coal fired power around the world, with the US seeing large reductions in coal use and even the assumed growth areas of China and India rapidly turning away from coal. There has been a corresponding dramatic collapse in the market capitalization of coal companies. All this will make the task of closing 1,000 plants easier.

RATION ELECTRICITY, GET DRESSED
FOR THE WAR, AND RAPIDLY DRIVE EFFICIENCY

In response to lower power supply, launch an urgent efficiency campaign matched with power rationing. Include a global campaign to change the temperature by one to two degrees centigrade in all temperature-controlled buildings (increase/decrease according to season). Make this part of the "war effort" as a public engagement technique, with large immediate power savings. On the back of this, launch an urgent mass retrofit program, including insulating walls and ceilings, installing efficient lighting and appliances, solar hot water, and so on across both residential and commercial buildings. This would have significant short-term job creation impacts.

RETROFIT ONE THOUSAND COAL POWER
PLANTS WITH CARBON CAPTURE AND STORAGE

Install CCS capacity on one thousand of the remaining power plants. This huge investment would be much simpler through international standardization. The CCS technology will also be needed for removal of CO_2 from the atmosphere later in the One Degree War (generating power using biomass and sequestrating the CO_2). CCS is not yet commercially viable and will require heavy government intervention. However, Randers strongly believes CCS will be mandated because it is a simple, albeit expensive, way of reducing greenhouse gas emissions, whereas I'm more skeptical. It's not important at this stage, as all technologies will develop and actions taken will adapt accordingly.

Update: CCS has continued to struggle to gain traction and it seems increasingly likely it may fail to ever do so with the price of renewables falling so fast. The wild card here is China, which has enough economic clout to change the global market for CCS. I personally remain skeptical but would be happy if proven wrong!

ERECT A WIND TURBINE OR SOLAR PLANT IN EVERY TOWN

Build in every town of one thousand inhabitants or more at least one wind turbine, solar thermal or solar photovoltaic (PV) plant instead. Beyond the CO_2 and renewable technology acceleration benefits, this would have the powerful impact of giving most people in the world a tangible physical connection to the "war effort."

> *Update: There has been an incredible boom in solar since 2007 but what we missed was the way this has become distributed, rooftop systems rather than even local power stations. This is driving massive disruption in the utility markets, where their whole business model is under threat from what is now referred to as the "death spiral" of centralized utilities. It may be better to change this task to rolling out more rooftop systems rather than power stations.*

CREATE HUGE WIND AND SOLAR FARMS IN SUITABLE LOCATIONS

Launch a massive renewable energy program focused primarily on concentrated solar thermal, solar PV, and wind power – on land and offshore. Given the urgency, the initial focus will need to be on those areas with most short-term potential for mass rollout, with finance supported by global agreement. On a global scale, various studies have shown how we could move to a 100 percent renewable energy system relatively rapidly. A recent global study showed how this could be achieved by 2030 with full baseload coverage. Of particular interest is that it concluded it would actually be cheaper than fossil fuels and nuclear power, due to the considerable efficiencies inherent in an energy system based on renewable generation and electricity use. All such modeling exercises are problematic and subject to controversy, but there is certainly massive potential in renewables with a war effort-type approach.

> *Update: Despite the boom referred to above in distributed systems, a need will remain for some central power stations and these will have to be from renewable sources. The wildcard has been the drop in demand for power in developed countries meaning that as more solar and wind gets installed there is an oversupply of generation capacity. It will require tough policy decisions to deal with this as it requires closing old dirty plants.*

LET NO WASTE GO TO WASTE

Ensure that all used materials are recycled and reused, at the very least to recover the embedded energy. To force this, limit production of virgin aluminum, cement, iron, plastics, and forest products – possibly through international agreements to restrict their use through higher prices or a special global emissions tax on virgin materials. Drive public recycling as part of the war effort (there are good examples here also from World War II, where mass public recycling drives focused on key materials).

RATION USE OF DIRTY CARS TO CUT TRANSPORT EMISSIONS BY 50 PERCENT

Launch large-scale replacement of fossil-fuel cars with chargeable electric vehicles – running on climate-neutral power – along with a massive boost in fuel-efficiency standards, bans on gas-guzzlers, and greater use of hybrid cars. Public repurchase and destruction of the most inefficient vehicles ("cash for clunkers" schemes) may help speed the transition and help address equity issues. Given the time it will take to scale up production, there will need to be rationing of the purchase of gasoline and diesel and other restrictions on their use such as special speed limits on fossil-fuel cars. Such restrictive measures would help drive acceptance of electric and efficient vehicles that would be free of such controls – the fast electric car can wave as it passes the old gas guzzler on the freeway!

In World War II, fuel in the United States was rationed at four gallons (per vehicle per week), then reduced to three gallons, and finally reduced in 1944 to two gallons. Alongside this, a national 35 mph speed limit was imposed, and anyone breaking the limit risked losing his fuel and tire rations. The government ran marketing campaigns to support these measures, such as advertisements asking, "Is this trip necessary?" and education campaigns on "how to spend a weekend without a car." It seems there were early-day environmentalists at the US Defense Department!

Update: Despite being earlier dismissed as not being viable by traditional auto companies, electric cars are going from strength to strength, driven by disruptive players like Tesla. The old auto companies are now playing catch-up. This is a key development as government is much more likely to act when a technology has started to move to scale production in the mainstream market and needs boosting rather than having to create a new market.

So a rapid, emergency response, which involved strong policy to get people into cars that were cheaper to run, fun to drive and cleaned up urban air would not be such a challenge – especially in countries with severe air pollution like China and India.

PREPARE FOR BIO POWER WITH CCS

Interestingly, the 1C-war may not see a large increase in the use of biofuels for land transport (not even second-generation fuels made from cellulose). It seems better for the climate to grow the cellulose and burn it in power stations with CCS, thereby removing CO_2 from the atmosphere while making power and heat. For this reason, boosting cellulose production (in plantations and elsewhere) will be key.

Update: Biofuels have been roundly condemned for their impact on global food prices. They have been further criticized for not really contributing to CO_2 reduction. This remains that industry's challenge but it is notable that 2nd generation biofuel producers have recently come on board arguing they can deliver dramatic "wells to wheels" CO_2 reduction using agricultural waste. Time will tell if they can do so at scale while managing the food price impact. So this may change the biofuels for land transport equation.

STRAND HALF OF THE WORLD'S AIRCRAFT

Reduce airplane capacity by a linear 10 percent per year through regulatory intervention and pricing to achieve a 50 percent reduction in airline emissions by the end of year 5. This will force the rapid development of biofuels for aircraft because of the commercial imperative to do so and force a cultural shift to electronic communication and away from frivolous air travel.

CAPTURE OR BURN METHANE

Put in place a global program to ensure that a significant proportion of the methane from agricultural production and landfills are either captured for energy purposes or at least burned to reduce the warming effect of that methane by a factor of 23.

MOVE AWAY FROM CLIMATE-UNFRIENDLY PROTEIN

Move society toward a diet with much less climate-unfriendly meat – through public education backed by legislation and pricing. This should not be against meat in particular, but against the associated emissions, so that preference is given to protein produced with lower emissions. There are large differences among protein types – emissions differ from soy, chicken, pork, and beef (and within beef, from grass vs. grain fed, particularly noting the emerging science that cattle grazed in certain ways can dramatically increase soil carbon). Therefore science based policy should be established to encourage the most impactful behavior change and for meat to be rated CO_2e/kg and priced accordingly. We note that the US government ran an effective "meat-free Tuesday" campaign during World War II. There is now already a community-based Meatless Monday campaign.

Update: It remains interesting how little real attention is directed towards this issue in the mainstream debate given the very high level of impact animal protein has. In hindsight one thing Randers and I may have paid insufficient attention to, given the context of the One Degree War Plan, was the relatively much greater importance of action on animal protein in an emergency response. Because of the way methane *behaves vs. CO_2 in the atmosphere (greater impact but with a much shorter half life) a genuine emergency response would strongly prioritize methane and therefore meat consumption, particularly cattle and sheep, as well as dairy. For example if you consider the impact of action just looking at the first 20 years, methane is 80 times as important as CO_2 in its warming impact. So in a true global emergency, the actions proposed above are still right but would be of greater importance.*

BIND 1 GIGATON OF CO_2 IN THE SOIL

Develop and introduce agricultural methods that reduce greenhouse gas emissions from agriculture and maximize soil carbon. This will require significant changes in farm technology and farmer psychology, and we are unlikely to get far during the first years. But the effort should be started immediately in preparation for the large-scale binding of carbon in forestry and agriculture that will be necessary from year 5 onward, in order to remove CO_2 from the atmosphere over the rest of the century. In both cases, the object will be to grow as much plant material as possible and ensure that the bound carbon ends in the soil or in subsurface storage, not back in the atmosphere. Currently, global forests bind some 3 $GtCO_2e/yr$. Hopefully – through the use of fast-growing tropical plantations, supplemented with industrial growth of algae – we could achieve the binding (and safe storage) of some 6 $GtCO_2e/yr$ from forestry and agriculture combined in future decades.

> *Update. This is an area I believe holds great potential – though it still lacks extensive enough scientific research to establish the scale of this potential. The vast scale of grazing agriculture and the possibility, with changes to practices, of changing soil carbon levels while increasing food production per hectare presents an enticing opportunity if it can be proven. Given the above comments on meat's impacts it deserves a great deal of focus.*

LAUNCH A GOVERNMENT AND COMMUNITY-LED "SHOP LESS, LIVE MORE" CAMPAIGN

In order to free up finance, manufacturing capacity, and resources for critical war effort activities, a large-scale campaign to reduce carbon intensive consumption, or at least stabilize it, would be of great help. This will align well with the general need to shift the economy away from carbon-intensive activities toward climate-friendly experiences. We would propose a bottom-up and top-down campaign to highlight the quality-of-life benefits of low-carbon lives with less stuff.

Update: This is a topic that has had the attention of scholars and philosophers for a very long time, focused not so much on environmental impact but rather on the impacts on enhancing human well-being. It is likely that in a crisis driven ultimately by consumption, this approach will gain a great deal of focus and may even become a mainstream idea. It is explored further in Chapter 5 by Per Espen Stoknes.

CONCLUSIONS

The full plan (Randers and Gilding 2010),[3] provides further details on these and other actions that would be required. These include how we could raise $2.5 trillion per year by year 5 via a global carbon tax and how this could be used to finance the measures required to compensate the poor, reduce disruption, and create the new industries and employment required. We also cover the types of multinational decision-making bodies that would be required, including a Climate War Command, and more detail on the actions required after the first five-year war, including major reversible global geo-engineering projects to reflect sunlight and remove CO_2 from the atmosphere and stabilize the global climate.

While all these actions may seem draconian or "unrealistic" by the standards of today's debate, they will seem far less so when society moves to a war footing and a focus on "what is necessary." Once more, World War II demonstrated that seemingly unachievable actions quickly became normal when delivered in the context of a war effort. In that case they ranged across dramatic increases in the level of taxation, the direction by government of manufacturing, and engagement campaigns to drive shifts in public behavior. So to emphasize this key point: the plan concludes that the challenge is not to find appropriate actions or to imagine how we could implement them. The challenge is to make the decision to act on the problem.

Professor Randers and I do not argue that we have the right plan or have defined all the right actions. What we sought to establish is that if society decided to fix climate change we can see through our plan that quite extraordinary reductions and management measures are practical and achievable and furthermore, that the economic cost would be considerably less than unchecked climate change.

Of course, there will be significant disruption as old industries are closed, as well as dislocations as people are moved on to new economic activity. But in a real war, such losses are caused by the decision to go to war. In our case,

---

3   Available from my website http://paulgilding.com/2009/11/06/cc20091106-odw-launch/

losses would occur anyway, because climate change will inevitably drive the collapse of the economy if strong action is not taken.

The exciting thing about such a plan is that, unlike in a real war, deciding to launch the One Degree War doesn't cost any lives. Instead it saves millions of them. It doesn't shift economic resources onto wasteful though necessary activities; it redirects them to build exciting new industries that will enhance the quality of life for the people of all countries involved. It doesn't waste a generation of youth and leave the survivors traumatized; it educates a generation in the technologies of the future and drives productive innovation that builds new companies and industries.

It is a war we have no choice but to fight with great benefits to be gained from its declaration.

BIBLIOGRAPHY

Davis, S.J., Caldeira, K. and Matthews, H.D. (2010). Future CO_2 emissions and climate change from existing energy infrastructure. *Science, 329*(5997), 1330–1333.

Juniper, T. (2014). What If Winston Churchill Were Leading the Fight Against Climate Change?, *Yes! Magazine,* http://www.yesmagazine. org/issues/how-to-eat-like-our-lives-depend-on-it/winston-churchill-leading-fight-climate-change

Matthews, H.D. and Weaver, A.J. (2010). Committed Climate Warming, *Nature Geoscience, 3*(3), 142–143.

Meadows, D.H., Meadows, D.L., Randers, J. and Behrens, W.W. (1972). *The Limits to Growth*. Universe Books: New York.

Meadows, D., Randers, J. and Meadows, D. (2004). *Limits to Growth: the 30-year Update*. White River Junction, Vt.: Chelsea Green Publishing.

Randers, J. and Gilding, P. (2010). The one degree war plan. *Journal of Global Responsibility, 1*(1), 170–188.

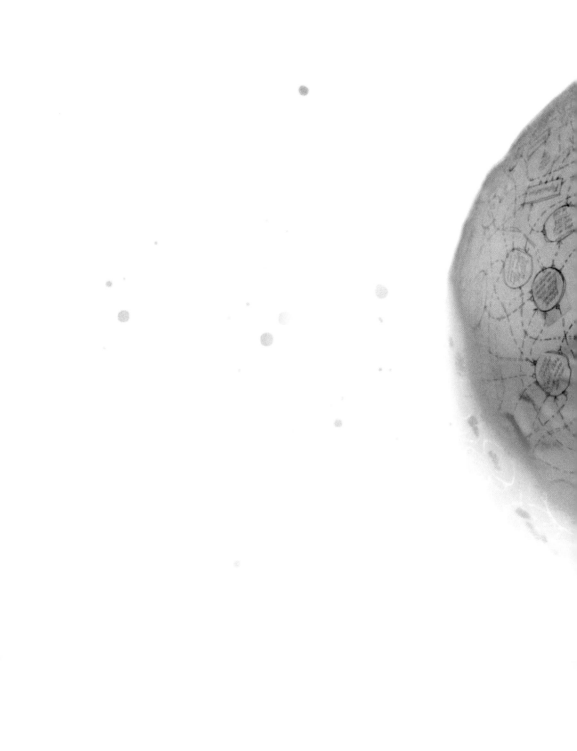

CHAPTER 3

Measures of Biodiversity – Living Planet Index and Nature Index as Tools for Activism

By Iulie Aslaksen, Per Arild Garnåsjordet,
Jonathan Loh, Kristin Thorsrud Teien[1]

---

1 The chapter builds on earlier cooperation with and contributions from Jane McDonald (University of Queensland, Australia) and Peter Cosier (Wentworth Group of Concerned Scientists, Australia) and we would like to acknowledge their contributions to the chapter.

INTRODUCTION: SHORT-TERMISM DESTROYING BIODIVERSITY

Economic development and rapid population growth over the last century have led to increased pressure on ecosystems and biodiversity. Improved indicators for measuring biodiversity loss are therefore needed worldwide, as well as better biodiversity monitoring giving data to indexes and accounting frameworks. With the increased awareness of environmental problems, numerous attempts have been made to develop aggregated measures for biodiversity. An early measure for biodiversity was the Living Planet Index (Loh et al. 1998). It was developed by the World Wildlife Fund (WWF) when Jorgen Randers was Deputy Director General, yet another contribution by Randers in his work for developing scientific knowledge for environmental policy and activism. The Living Planet Index is a widely used tool for communication of biodiversity loss. The most recent Living Planet Report 2014 documents a 52% decline between 1970 and 2010 of the species included in the Living Planet Index (LPI, WWF 2014).

The Convention on Biological Diversity (CBD 1992) suggests this definition of biodiversity: "Biological diversity means the variability among living organisms from all sources including, inter alia, terrestrial, marine and aquatic ecosystems and the ecological complexes which they are part of; this includes diversity within species, between species and of ecosystems." CBD identifies five main direct threats to biodiversity globally: habitat loss and degradation, invasive alien species, pollution and nutrient load, overexploitation and unsustainable use, and climate change. According to the Millennium Ecosystem Assessment (MEA 2005), changes in agriculture, forestry and physical developments of infrastructure, industry and urban areas are the greatest threats to biodiversity in terrestrial ecosystems, while fisheries are the greatest threats to marine ecosystems. In aquatic ecosystems, changes in water management, invasive species and pollution, in particular eutrophication from excessive nutrients, are the greatest threats.

The 2010 Conference of the Parties to the Convention on Biological Diversity (COP 10) in Nagoya, Japan set an ambitious target: "By 2020, the rate of loss of all natural habitats, including forests, is at least halved and where feasible brought close to zero, and degradation and fragmentation is significantly reduced" (Target 5). And target number 2 sets the ambition for accounting the values of biodiversity: "By 2020, at the latest, biodiversity values have been integrated into national and local development and poverty reduction strategies and planning processes and are being incorporated into national accounting, as appropriate, and reporting systems" (UNEP/CBD 2010).

TEXTBOX 3.1: Reasons to reconsider the role of biodiversity in environmental policy, in addition to an ethical responsibility to preserve biodiversity for its own sake.

Biodiversity is:

▶ *An intrinsic part of the natural world that ought to be protected (Nash 1989)*
▶ *A responsibility of humanity (WRI 2005)*
▶ *Critical to achieving sustainability (MEA 2005)*
▶ *The essential foundations upon which humanity depends (CBD 1992)*
▶ *Representative of conservation as a whole (MEA 2005)*
▶ *Esthetic qualities (Ehrlich and Ehrlich 1981)*
▶ *Biodiversity is a "good" (Mace et al. 2012)*
▶ *Essential for the functioning of ecosystems that underpin the provisioning of ecosystems that affect human well-being (MEA 2005)*
▶ *Insurance against future unknown threats (MEA 2005)*
▶ *An unknown potential source of benefits, e.g. pharmaceuticals (MEA 2005)*
▶ *An indicator of ecosystem condition (Karr 1991)*
▶ *A measure of only species extinctions (MEA 2005)*
▶ *A measure of all of biology (Sarkar 2005)*
▶ *A major factor affecting ecosystem stability (May 1975)*
▶ *Correlated to productivity (more diverse communities are more productive) (Darwin 1872)*
▶ *An input influencing many ecosystem properties (Tilman and Downing 1994, UK NEA 2011)*
▶ *An important element in the functioning of ecosystems (MEA 2005)*
▶ *Critical to the viability of indigenous communities (WWF 1997)*
▶ *A unique and irreplaceable part of our world (MEA 2005)*
▶ *Providing incalculable benefits of genetic variability that people everywhere use daily and depend upon (McAfee 1999)*
▶ *Inherent in all ecosystems, not an entity that can be separated (MEA 2005)*
▶ *Contributes to security, resiliency, social relations, health and freedom of choices and actions (MEA 2005)*
▶ *Biodiversity is synonymous with ecosystem services (TEEB 2010)*
▶ *Supports cultural value (MEA 2005)*

Source: (McDonald 2011).

The decline of particular species and habitats has drawn both scientific and public attention. And this increasing recognition of biodiversity loss has strengthened political initiatives for biodiversity conservation. But more fundamentally it has also resulted in closer examination of why biodiversity is important, how it is an essential foundation for life on earth and what role it plays in human civilization (McDonald 2011). What will global biodiversity look like in the future? What are possible political responses to the short-termism destroying biodiversity? To what extent can biodiversity measurement be a tool for policy for a planet under pressure?

3.1   THE LIVING PLANET INDEX

The LPI began in 1997 as a WWF project to develop a measure of the changing state of the world's biodiversity over time. Work on the LPI started in collaboration with the UNEP World Conservation Monitoring Centre (WCMC) and, since 2005, continued with the Zoological Society of London (ZSL). The first index was published in the WWF Living Planet Report 1998 (Loh et al. 1998) and has been updated subsequently (see e.g. Loh, 2002, Loh et al. 2005, WWF 2014). The LPI aims to measure average trends in populations of vertebrate species from around the world since 1970. The index is currently based on more than 10 000 population time series for over 3 000 mammal, bird, reptile, amphibian and fish species. The initial aim was to make the LPI as comprehensive and representative as possible with respect to taxonomic vertebrate class, biogeographic realm, biome, and threat status. The restriction of the index to vertebrate animals is for reasons of data availability: on a global scale, relatively few time-series data for invertebrate or plant populations exist, and those available come from geographically restricted locations. The LPI is a measure of global biodiversity only as far as trends in vertebrate species populations are representative of wider trends in all species, genes and ecosystems. The principle is species abundance: the more individuals, the better are conditions. All species contribute equally in the calculations of the index.

The Living Planet index has since inspired numerous other approaches to biodiversity measurement, including the official Nature Index for Norway.

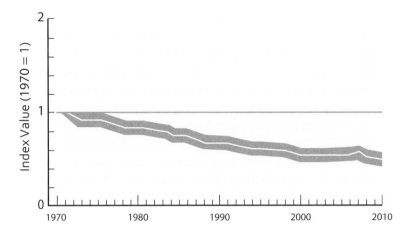

FIGURE 3.1:   The Living Planet Index. (Source: WWF Living Planet Report 2014)

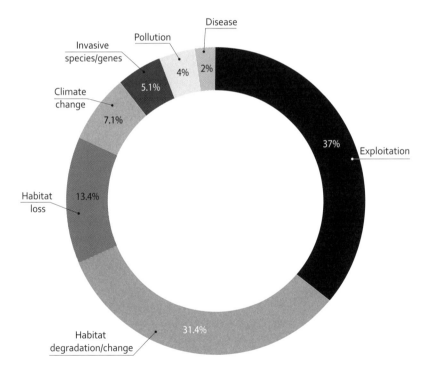

FIGURE 3.2:   Primary threats to LPI populations. (Source: WWF Living Planet Report 2014)

3.2   A LIVING PLANET INDEX FOR NORWAY

In 2005, WWF in Norway made a pioneering contribution[2] to the development of a biodiversity index for Norway, based on the LPI. This index was called the Nature Index for Norway, a term that later became the official *Nature Index for Norway*. The purpose was to achieve a "pulse meter" for nature. This was an ambitious, but necessary, goal. What is not measured often gets overlooked in management and policy. Clear targets for biodiversity are needed in environmental policy. The Norwegian parliament and the EU decided in 2003 to adopt the goal of bringing biodiversity loss to a halt by 2010. The LPI data for Norway included vertebrates and large crustaceans, with 568 long time series for species in sea, land and freshwater (marine, terrestrial, and aquatic habitats). They were mostly common species, many with large importance economically and culturally, as well as ecologically (Teien 2005a).

The Living Planet Index for Norway showed that nature was going downhill: The included species had on average a reduction in abundance of 35 percent in less than 30 years, from 1975 to 2003. They were near the top of the food web and thus probably reflected the state further down the food web. The ecological and economic consequences in the long run are hard to envision. There was a dramatic decline for the ocean species, including 15 fish species, 21 bird species, three marine mammals, shrimp and lobster. Stocks of marine fish and crustaceans were found to be reduced by 50 percent over the period. The northern Norwegian populations of the sea birds guillemot and lesser black-backed gull had a particularly large decline. Many populations of species related to freshwater and mires and wetlands were reduced in numbers, on average by 50 percent.

Economic activity and public policy contributed in various ways to shaping this decline: The halving of marine fish stocks was probably mainly caused by overharvesting. Loss of freshwater fish stocks was first and foremost due to acidification from industry. The Living Planet Index for Norway also showed examples of management actions with effects – populations of trout improved

---

2   One of the authors of the chapter, Kristin Thorsrud Teien, would like to add a personal note of gratitude to Jorgen Randers. Before her affiliation with the Norwegian Ministry of Climate and Environment, while she was employed by WWF Norway, she took the lead in developing, for the first time, a Living Planet Index for Norway. The Living Planet Index for Norway was a novel approach to synthesizing ecological knowledge, based on input and data from a large number of different sources. The development of this biodiversity index for Norway would not have been possible without Randers as a source of inspiration and sparring partner. Randers never failed in his enthusiasm for this pioneering effort to develop trends for Norwegian nature. Kristin thanks Randers for his long-lasting support and friendship during this work and their shared love for nature.

considerably after adding lime to rivers and lakes to counteract acidification. The Living Planet Index for Norway brought more attention to the lack of data from surveying and monitoring. Species cannot be managed without updated knowledge about their development (Teien 2005b).

Developing the Living Planet Index for Norway was a remarkable example of scientific activism from WWF. The index became a wake-up call for society. The trends reported aroused considerable concern for environmental management and spurred the decision to develop an official Nature Index for Norway, mandated by the Government in 2005.

3.3   THE NATURE INDEX FOR NORWAY

The Nature Index for Norway is a flexible and comprehensive framework for integrated biodiversity measurement that builds on and further extends internationally well-known biodiversity indices, such as the Living Planet Index, the Natural Capital Index, the GLOBIO index and the Biological Intactness Index (Loh et al. 2005, ten Brink et al. 2002, Alkemade et al. 2009, Scholes and Biggs 2005). The aim of the Nature Index, initiated by the Norwegian government, was to provide a tool for measurement of the state and trends of biodiversity, in order to assist environmental managers and policymakers in setting biodiversity policy objectives and monitoring priorities (Nybø (ed.) 2010, Certain and Skarpaas et al. 2011, Nybø et al. 2012).

The political purpose of the Nature Index is to give an overall picture of the state of Norway's biodiversity. Therefore it was built on the following principles: All the major taxonomic groups are represented, both common and rare species are represented, indicators should be complementary with regard to their response to anthropogenic pressures, keystone species should be included when possible, and a wide variety of ecosystems and habitat should be represented by the indicator set.

In total 309 biodiversity indicators were used in the Nature Index. The weighting system was designed to control for biases arising from the over- or underrepresentation of certain taxa and functional groups (trophic level). The Nature Index was established through an extensive cooperation between experts from leading research institutions, where 125 scientists participated in defining the criteria, selecting the biodiversity indicators, and entering the data, consisting of monitoring data, model-based data and expert judgment. The inclusion of expert judgment makes it possible to apply the Nature Index framework also in data-poor countries or regions.

The Nature Index for 2010 showed that the state of biodiversity was highest relative to the reference states in mountains, ocean, coast, and freshwater,

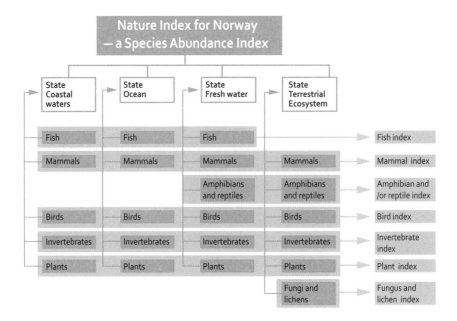

FIGURE 3.3:   Nature Index for Norway as framework for aggregating indicators distributed over ecosystems. For simplicity, terrestrial ecosystems, separated into forest, open lowland, mires and wetlands, and mountains, are here shown as one ecosystem. Source: Nybø et al. (2012). Reprinted with permission.

intermediate for mires and wetlands, while open lowlands and forests had the lowest values. The difference between the Nature Index and the Living Planet Index for marine ecosystems reflects different methods: The LPI for Norway only reported trends, while the Nature Index reports current state relative to a reference state, expressed by the best available knowledge of an intact ecosystem. Another difference is that the species that are included differ between the two indexes. The low states for the terrestrial ecosystems reflect impacts of forestry and modern agriculture. For forests, also the low populations of big carnivores, with the related too high densities of wild herbivores such as moose and deer in many areas, play a major role in the resulting low state of biodiversity. The situation in open lowlands gives the largest cause for concern, as the abandonment of traditional agricultural practices and increasing fertilization and re-growth of forests threaten the characteristic biodiversity of the traditionally managed cultural landscape.

A flexible decomposition of the Nature Index makes it possible to show what types of changes are behind the aggregate picture. Thematic indices can be calculated along the dimensions of species groups, taxonomic groups or major ecosystems but also for species being sensitive to defined pressure factors. The Nature index can be aggregated or disaggregated to address specific management themes.

3.4 COMPOSITE BIODIVERSITY INDICATORS

The complexity of biodiversity and its relationships to ecosystems implies that selecting only a few indicators will not give a reasonably complete picture of this multivariate concept. Thus, a "system of indicators" is needed. Different types of indicator systems may then be used for communication and policy purposes, representing the perspectives and values for different stakeholders. As in any aggregation and compilation of scientific information of high complexity, this process is not "value-neutral," and every step entails normative choices and evaluations (ten Brink 2006).

A composite biodiversity indicator may be seen as a "tip of the iceberg," where the underlying information serves to support the interpretation of the aggregate index. Composite indicators can be seen as data-based narratives. They implicitly describe relationships between indicators belonging to different dimensions. While including analytic elements, they are still a constructed view of reality, and thus sit somewhere between analysis and advocacy, integrating science and activism (Saltelli 2007).

According to the authors of the Stiglitz report (Stiglitz 2009, p. 65) the normative element of composite indicators is often not explicit enough. This is particularly relevant for biodiversity issues reflecting the interests of different stakeholders in society. Composite indicators are often based on pragmatic choices of data availability, and also on certain assumptions of what elements are the most important to represent and which pressure factors are operative.

There are many different types of biodiversity indicators, including internationally well-known indicators of abundance and richness of species, statistical measures of ecological complexity, and indicators including the concept of reference conditions to measure the quantity and quality of an ecosystem compared to an intact condition. By comparing indicator levels to a common reference or baseline (as in many of the composite indices) indicators can be made comparable across countries (ten Brink 2006).

3.5 WHY ACCOUNT FOR BIODIVERSITY?

Biodiversity indicators and indices need to be integrated in frameworks that are related to policy targets and instruments. The Living Planet Index was the foundation for subsequent biodiversity measures of varying complexity. The Nature Index for Norway has a complex framework that responds to the call for integrated biodiversity measurement. Yet it is not developed completely in terms of formulation of policy and management targets.

A challenge for biodiversity policy is that conservation organizations are called upon to be responsive to a number of economic, social and ethical concerns beyond that of the conservation of biodiversity (Brauer 2005, Laikre et al. 2008, Ahlroth and Kotiaho 2009). In spite of numerous scientific approaches to address biodiversity decline, only a few have been able to synthesize this information to give an overview of the overall global trends (e.g., Walpole et al. 2009). For establishing accounts for biodiversity, a careful consideration of the concept of biodiversity and its relationship to ecosystem functioning is needed, with a step-by-step process to determine what components of biodiversity are being accounted for and why. This will help explain how different aspects and values of biodiversity are accounted for, and what measures are appropriate for different purposes, emphasizing that (1) biodiversity is an indicator of the ecosystem condition, (2) biodiversity is an intrinsic value of the natural world, and (3) biodiversity has value for human benefit, as an ecosystem service (intermediate or final).

The ecosystem services valuation approach represents a trend of increasing focus on monetary valuation of biodiversity and the ecosystem services it gives rise to (TEEB 2010). But Mace et al. (2012) has argued that the relationship between biodiversity and ecosystem services is confused and is damaging efforts to create coherent policy. Numerous objections have also been raised against relying on monetary valuation of ecosystem services as a primary policy tool (Spash 2008). Biodiversity measurement generally concentrates on the diversity of species of flora and fauna, as well as the abundance, ecological function, community composition, and spatial distribution of these species. These factors are of crucial importance for the ecosystem services providing economic value. Economic policy is focused on improving human living standards by continually *expanding* the value of the flows of goods and services. On the other hand, environmental policy is about *maintaining* the stock (condition) of natural capital, including ecosystems, so that they continue to provide services for humanity into the future (Cosier 2011). The main argument for biodiversity accounting is to make it possible to balance and/or integrate the two policy purposes.

We may also envision the accounting of ecosystem services as a tool for improved environmental management on a local level. This has recently been pioneered by Australia: Ecosystem services (bundles of flows measured as Environmental Benefit Indices) are proposed to distribute stewardship funds for the Victoria Department of Sustainability and Environment in Australia (Eigenraam 2011). Examples of selected components of biodiversity include the Habitat Hectares metric, which has been used as a surrogate for biodiversity in Victoria (Parkes et al. 2003). This is an index of vegetation condition, extent and connectivity relative to an "undisturbed" reference state (baseline). In Australia, recent approaches to environmental accounting have been further developed on a regional and national scale (Sbrocchi 2013, Wentworth Group 2014).

3.6 EXPERIMENTAL ECOSYSTEM ACCOUNTING

A system of *Experimental Ecosystem Accounting (EEA),* aiming to integrate ecological valuation as the basis for economic valuation, is currently being developed by the United Nations (2013). It takes into account ecological and biophysical biodiversity measurement, in order to represent biodiversity in environmental accounting. A core concept in EEA is ecosystem capacity – the capacity to sustain over time the delivery of a "basket" or "bundle" of multiple ecosystem services from a particular ecosystem at a given spatial level. *Ecosystem capital* as an economic concept is defined as the present value of expected ecosystem services, as envisioned today (TEEB 2010).

Ecosystem capacity represents a much larger potential, taking into account that future trade-offs between "baskets" or "bundles" of multiple ecosystem services may require a larger capacity for ecosystem services than the current ecosystem capital. Hence, valuation of a basket of multiple ecosystem services cannot be based on current economic valuation. If future decision-making will prioritize biodiversity, carbon storage and recreation in forests, in addition to forestry, the asset values of ecosystem capacity will be much higher than what is reflected by ecosystem capital based on today's timber and recreational entertainment prices (see Norman, this volume).

The use and valuation of multiple ecosystem services need to be seen together, so that use of one ecosystem service (e.g. provisioning) will not undermine the potential to maintain other ecosystem services. This approach suggests a close connection between valuation of ecosystem services and assessment of ecosystem condition. In order to assess the ecosystem capacity, it must be taken into account that a reduction in biodiversity may reduce the capacity to deliver different ecosystem services in different ways. Loss of supporting and regulating ecosystem services may be less visible to society as long as the provisioning of services continues.

The UN Experimental Ecosystem Accounting system explicitly considers how the potential for ecosystem services depends on biodiversity and other bio-physical conditions, such as the landscape dimension, and how ecosystem effects change the potential for future ecosystem services. The system is based on a geographical grid of basic spatial units, which may be aggregated to a set of ecological accounting units (e.g. administrative units or other delineations, such as watersheds), each comprising several types of ecosystems.

The experimental ecosystem accounting system combines the use of ecosystem services and the impact on the state of ecosystems in sets of tables for each ecological accounting unit. Data from various sources are integrated in tables showing the extent and ecological state, and change in state, for each type of ecosystem, and use of ecosystem services from these ecosystems. These tables comprise a consistent area-based ecosystem accounting framework. They show how human impacts related to the use of different ecosystem services from different ecosystems in a given area will change the extent and state of the ecosystems and thus the potential for future ecosystem services. The main advantage of this framework is that it directly relates the impact of human activities through use of ecosystem services, land use change, climate change and pollution, to the potential for future ecosystem services – through the impact on biodiversity and extent of state of ecosystems – integrated into a common area-based accounting system.

3.7   THE USEFULNESS OF BIODIVERSITY ACCOUNTS FOR POLICY

Some of the main challenges in biodiversity policy include information and communication of changes in biodiversity, a process that may require changes in the indicator set and the data-based "narratives" that are formulated. Biodiversity indicators can be presented as a gauge of observed changes without specifying what the causes or drivers might be. However, for the interpretation of trends or developments over time to be meaningful, it has to be stated whether changes are positive or negative relative to a reference state (baseline).

If the underlying causes or drivers for biodiversity change can be substantiated, the narratives are strengthened; for instance "Over-fishing of this species also leads to decline of another, etc." This process does not involve scientists alone, but also the public, interest groups, and policy makers. Active adaptive management may be formulated based on cost-benefit analyses or multi-criteria analyses that make it possible to evaluate the cost-effectiveness in environmental management and restoration (Cosier 2011). Communication of biodiversity as a policy issue can be enhanced by adopting a participatory approach including both the general public as well as national and local stakeholders (Garnåsjordet et al. 2012).

Biodiversity information has to be communicated in an easily understandable form. An indicator should therefore be like a temperature gauge on the dashboard of a car, which shows the driver that the engine is performing effectively. This signal is easy to understand and allows for immediate action to be taken. The indicator set, as a whole, in the dashboard is carefully designed and selected to provide the driver with information that allows for driving safely. Speed, distance to the target, fuel level, fuel consumption, and direction cannot be added into one indicator, but need to interpreted as complementary elements of information. This is recommended by the Stiglitz commission:

> *The assessment of sustainability is complementary to the question of current well-being or economic performance, and must be examined separately. This may sound trivial and yet it deserves emphasis, because some existing approaches fail to adopt this principle, leading to potentially confusing messages. For instance, confusion may arise when one tries to combine current well-being and sustainability into a single indicator. To take an analogy, when driving a car, a meter that added up into one single number the current speed of the vehicle and the remaining level of gasoline would not be of any help to the driver. Both pieces of information are critical and need to be displayed in distinct, clearly visible areas of the dashboard. (Stiglitz et al. 2009, p. 17)*

The Convention on Biological Diversity (CBD) has suggested that policymakers address four simple questions when considering biodiversity: What is changing, why is it changing, why is it important and what can be done (ten Brink 2006)? These key questions relate directly to the policy cycle and feedback principles. Effective biodiversity management is only possible if the following conditions are met: (1) There are verifiable policy targets for biodiversity. (2) Timely and sufficient knowledge of the current and projected states is established as well as progress made towards the targets. (3) It is possible to make corrections. Accounting with a measure of critical values for biodiversity facilitates the setting of limits for a "safe operating distance" for biodiversity critical values and identifies thresholds for policy targets (Rockström et al. 2009).

We can now formulate some important policy recommendations: (1) Any approach to accounting for biodiversity needs to take into account that biodiversity includes the variability of species and the ecosystems that the species are part of and recognize that ecosystems are a completely integrated part of biodiversity (Mace et al. 2012). (2) Careful consideration of the concept of biodiversity and its relationship to ecosystem functioning is recommended, with a step-by-step process to determine what components of biodiversity are being accounted for and why. (3) The approach to biodiversity accounting needs to consider how different aspects and values of biodiversity are accounted for, and what measures are appropriate for different purposes.

CONCLUSION:
TOWARD NEW FORMS OF SCIENCE-POLICY DIALOGUES

The importance of biodiversity as the very basis for life-supporting ecosystems provides an ethical imperative to give high priority to the protection of biodiversity (Heywood and Iriondo 2003). Biodiversity policy needs to be understood in a context of post-normal science. Post-normal science encompasses the extension of scientific practice into situations in which scientists take into account the intertwined relationships between facts and values, the possibility of catastrophic decision-stakes, the legitimate plurality of conflicting interests and ethical complexities, beyond what is usual in normal scientific practice (Funtowicz and Ravetz 1990). Biodiversity brings up "a range of urgent problems that require immediate attention but cannot be adequately addressed by current scientific knowledge or methods, relies heavily on practitioners who are not scientific experts, (an extended 'peer community'), where decisions made may have substantial repercussions regarding human lives and livelihoods, and in which laypersons from a range of backgrounds have a stake" (Francis and Goodman 2010).

Involvement and participation of stakeholders and citizens may contribute to improving the quality of the policy deliberations (Funtowicz and Strand 2011). Biodiversity indices can be applied to express political objectives or serve as input for different types of deliberations or communication processes in society. In the Nature Index for Norway, for instance, experts were asked about their assessments of future biodiversity trends, consequences of biodiversity loss, and the need for urgent management action (Aslaksen et al. 2012).

Framing biodiversity only in an economic policy context does not allow for the extension of the model of human agency in terms of more inclusive participation and deliberation. It does not help much that almost "everyone" is aware of how the materialistic lifestyle of Western societies is an increasingly serious threat to ecological balance and biodiversity – as long as policies to protect nature and biodiversity seem almost impossible to achieve within the current context of modern democratic capitalist societies. It seems clear today that the required policies will probably not be adopted in time to maintain biodiversity and ecological balance. The main reasons are clearly related to the short-termism – the short-term thinking – inherent in attitudes and actions on all levels of society, and the widespread resistance to public regulation.

The problem is not only short-termism. A focus on the long-term does not by itself address the question of how we should live today without postponing the difficult decisions. How do we act in a responsible way *today*? The short-termism of the modern world is portrayed by philosopher Hannah Arendt (1958) in *The Human Condition*, where she warns against the erosion of the community and solidarity of the public sphere, as a consequence of the relentless pursuit of economic interests. Her concept of *earth alienation* captures how the technological dominance of the modern age has led to the loss of nature and to a profound transformation of the natural environment as well as of our relationship to nature.

The dramatic threats to biodiversity and ecological balance are hardly reflected within mainstream economics – framed in a policy context of short-termism. The language of power is the language of conventional economics. Those who are in power also hold the power of definition. "Economists and policymakers speak the same language," says ecologist Ben ten Brink (2006). The immeasurable values and qualities of nature still remain invisible in a policy context that primarily seeks economic growth. On the global level, policies for biodiversity protection are often perceived as being in conflict with policies of food security for an increasing world population. Food security is largely framed in a policy context dominated by the agro-industrial complex, and the call for integration of agricultural and environmental policies is marginalized.

According to the report of the International Assessment of Agricultural Knowledge, Science and Technology for Development (IAASTD 2008), technological progress in intensified, "modern" agriculture has led to a huge reduction of biodiversity. But the seemingly "obsolete" systems of traditional agricultural production often show high performance if more comprehensive ecological and social criteria of performance are included. Many traditional forms also protect the genetic diversity of crops, and revitalize and empower local communities.

The Convention on Biological Diversity (CBD) expresses the commitment to achieve "a significant reduction of the current rate of biodiversity loss at the global, regional and national level as a contribution to poverty alleviation and to the benefit of all life on earth." Currently, the economic interests of the agro-industrial complex based on human and political short-termism pull us – blindly – in the opposite direction. Ecosystem accounting and biodiversity indices may help us – at least – to open our eyes.

The understanding of human identity in our culture is based on separation from nature, rather than by the system relationships between humans and nature. In contrast to this view, the renowned biologist Edward O. Wilson coined the term *biophilia* to describe the fundamental emotional affiliation of the human being to Nature. We are not only dependent on Nature for our physical needs, but also in a deeper sense, for emotional, mental and spiritual well-being. Wilson also pointed out the moral dimension: "The affiliation has a moral consequence: the more we come to understand other life forms, the more our learning expands to include their vast diversity, and the greater the value we will place on them, and inevitably, on ourselves" (Wilson 2006).

The economic and financial crisis of the Western world has led to an increasing lack of credibility and trust in the political and economic institutions of society. The struggle to reverse the current erosion of the public sphere, despite all the difficulties, is the way we understand the potential for improvement of the human condition and addressing the environmental threats.

Ecosystem management targets need to balance the prioritized economic use of a given ecosystem at a given spatial level with society's view of what is a sustainable capacity to deliver the "bundle" of future ecosystem services. This calls for new forms of dialogue. The deliberation on management targets for biodiversity of high value and vulnerability, the application of an experimental ecosystem accounting framework, and the understanding of the causes of biodiversity loss may contribute to the integration of economic and ecological approaches to ecosystem effects into the knowledge basis for biodiversity policy.

BIBLIOGRAPHY

Ahlroth, P. and Kotiaho, J. (2009). Route for political interests to weaken conservation. *Nature, 60*, 173.

Alkemade, R., van Oorschot, M., Miles, L., Nelleman, C., Bakkenes, M. and ten Brink, B. (2009). GLOBIO3: A Framework to Investigate Options for Reducing Global Terrestrial Biodiversity Loss. *Ecosystems, 12*(3), 374–390.

Arendt, H. (1958). *The Human Condition*, University of Chicago Press.

Aslaksen, I., Framstad, E., Garnåsjordet, P.A., Nybø, S. and Skarpaas, O. (2012). Knowledge gathering and communication on biodiversity: Developing the Nature Index for Norway. *Norsk Geografisk Tidsskrift – Norwegian Journal of Geography, 66*, 300–308.

Brauer, J. (2005). Establishing Indicators for Biodiversity. *Science, 308*(5723), 791–792.

Certain, G., Skarpaas, O., Bjerke, J-W., Framstad, E., Lindholm, M., Nilsen, J-E., Norderhaug, A., Oug, E., Pedersen, H-C., Schartau, A-K., van der Meeren, G.I., Aslaksen, I., Engen, S. Garnåsjordet, P.A., Kvaløy, P., Lillegård, M., Yoccoz, N.G. and Nybø, S. (2011). The Nature Index: A General Framework for Synthesizing Knowledge on the State of Biodiversity. *PloS ONE, 6*(4), e18930.

Convention on Biological Diversity (1992). https://www.cbd.int/doc/legal/cbd-en.pdf

Cosier, P. (2011). *Accounting for the condition of environmental assets*, in UN Committee of Experts on Environmental Accounting Technical Meeting on Ecosystem Accounts: London, 5–7 December 2011.

Darwin, C. (1872). *The origin of species*. Sixth London Edition ed. 1872, Chicago, Illinois: Thompson and Thomas.

Ehrlich, P.R. and Ehrlich, A. (1981). *Extinction:The Causes and Consequences of the Disappearance of Species,* New York: Random House.

Eigenraam, M., Vardon, M., Hasker, J., Stoneham, G. and Chua, J. (2011). *Valuation of ecosystem goods and services in Victoria, Australia.* Information paper for Expert Meeting on Ecosystem Accounts, London, 5–7 December 2011.

Francis, R.A. and Goodman, M.K. (2010). Post-normal science and the art of nature conservation. *Journal for Nature Conservation, 18*(2), 89–105.

Funtowicz, S. and Ravetz, J.R. (1990). *Uncertainty and Quality in Knowledge for Policy.* Dordrecht: Kluwer Academic Publisher.

Funtowicz, S. and Strand, R. (2011). Change and commitment: beyond risk and responsibility. *Journal of Risk Research, 14*, 995–1003.

Garnåsjordet, P.A., Aslaksen, I., Giampietro, M., Funtowicz, S. and Ericson, T. (2012). Sustainable Development Indicators: From Statistics to Policy. *Environmental Policy and Governance, 22*, 322–336.

Heywood, V.H. and Iriondo, J.M. (2003). Plant conservation: old problems, new perspectives. *Biological Conservation, 113*, 321–335.

IAASTD (2008). *International assessment of agricultural knowledge, science and technology for development*, Executive summary of the synthesis report.

Karr, J.R. (1991). Biological Integrity – a Long-Neglected Aspect of Water-Resource Management. *Ecological Applications, 1*(1), 66–84.

Laikre, L., Jonsson, B.-G., Ihse, M., Marissink, M., Gustavsson, A.-M., Ebenhard, T., Hagberg, L., Stål, P.-O., von Walter, S. and Wramner, P. (2008). Wanted: scientists in the CBD Process. *Conservation Biology, 22*(4), 814–818.

Loh, J. (2002). *Living Planet Index 2002.* Gland, Switzerland: World Wildlife Fund International.

Loh, J., Green, R.E., Ricketts, T., Lamoreux, J., Jenkins, M., Kapos, V. and Randers, J. (2005). The Living Planet Index: using species population time series to track trends in biodiversity. *Philosophical Transactions Royal Society B, 360*, 289–295.

Loh, J., Randers, J., MacGillivray, A., Kapos, V., Jenkins, M., Groombridge, B. and Cox, N. (1998). *Living Planet Report 1998.* Gland, Switzerland: WWF.

Mace, G.M., Norris, K. and Fitter, A.H. (2012). Biodiversity and ecosystem services: a multilayered relationship. *Trends in Ecology & Evolution, 27*(1), 19–26.

May, R.M. (1975). Stability in multi-species community models. *Math. Biosci, 12*, 59–79.

McAfee, K. (1999). Selling nature to save it: Biodiversity and green developmentalism. *Environment and Planning D: Society and Space, 17*, 133–154.

McDonald, J. (2011). *Key concepts for accounting for biodiversity.* Prepared for the Expert Meeting on Ecosystem Accounts. London, 5–7 December 2011.

Millennium Ecosystem Assessment (MEA) (2005). *Ecosystems and Human Well-Being – Biodiversity Synthesis.* Washington, D.C.: World Resources Institute.

Nash, R.F. (1989). *The rights of nature: a history of environmental ethics.* Madison, Wisconsin: University of Wisconsin Press.

Nybø, S. (red.) (2010). *Naturindeks for Norge 2010.* DN-utredning 3-2010. Direktoratet for naturforvaltning, Trondheim.

Nybø, S., Certain, G. and Skarpaas, O. (2012). The Nature Index – state and trends of biodiversity in Norway. *Norsk Geografisk Tidsskrift – Norwegian Journal of Geography, 66*, 241–249.

Parkes, D., Newell, G. and Cheal, D. (2003). Assessing the quality of native vegetation: The "habitat hectares" approach. *Ecological Management & Restoration, 4*, 29–38.

Rockström, J., Steffen, W., Noone, K., Persson, Å., Chapin, F.S. III, Lambin, E., Lenton, T.M., Scheffer, M., Folke, C., Schellnhuber, H., Nykvist, B., De Wit, C.A., Hughes, T., van der Leeuw, S., Rodhe, H., Sörlin, S., Snyder, P.K., Costanza, R., Svedin, U., Falkenmark, M., Karlberg, L., Corell, R.W., Fabry, V.J., Hansen, J., Walker, B., Liverman, D., Richardson, K., Crutzen, P. and Foley, J. (2009). A safe operating space for humanity. *Nature, 461*(2), 472–475.

Saltelli, A. (2007). Composite indicators between analysis and advocacy. *Social Indicators Research, 81*, 65–77.

Sarkar, S. (2005). *Biodiversity and Environmental Philosophy: An Introduction. Cambridge Studies in Philosophy and Biology.* New York: Cambridge University Press.

Sbrocchi, C. (2013). *Guidelines for Constructing Regional Scale Environmental Asset Condition Accounts: Quick Guide.* Sydney: Wentworth Group of Concerned Scientist.

Scholes, R.J. and R. Biggs (2005). A biodiversity intactness index. *Nature, 434*(7029), 45–49.

Spash, C.L. (2008). How Much is that Ecosystem in the Window? The One with the Bio-diverse Trail. *Environmental Values, 17*, 259–284.

Stiglitz J., Sen, A. and Fitoussi, J.P. (2009). *Report by the Commission on the Measurement of Economic Performance and Social Progress.* www.stiglitz-sen-fitoussi.fr

Teien, K.T. (2005a). *Utfor bakke med norsk natur. Naturindeks for Norge 2005.* WWF rapport.

Teien, K.T. (2005b). Dramatisk naturutvikling. *Aftenposten.* Kronikk. 12. september 2005.

ten Brink, B. (2006). *A Long-Term Biodiversity, Ecosystem and Awareness Research Network. Indicators as communication tools: an evolution towards composite indicators,* www.alter-net.info

ten Brink, B.J.E. and Tekelenburg, T. (2002). *Biodiversity: how much is left? The Natural Capital Index framework (NCI).* in RIVM report 402001014. Bilthoven.

The Economics of Ecosystems and Biodiversity (TEEB) (2010). *Mainstreaming the Economics of Nature: A synthesis of the approach, conclusions and recommendations of TEEB.*

Tilman, D. and Downing, J.A. (1994). Biodiversity and stability in grasslands. *Nature, 367*(6461), 363–365.

UK National Ecosystem Assessment (UK NEA) (2011). *Biodiversity in the Context of Ecosystem Services,* in *UK National Ecosystem Assessment.* Chapter 4.

UNEP/CBD (2010). *Annex Strategic Plan for Biodiversity 2011–2020 and the Aichi biodiversity Targets "Living in harmony with nature" Decision X/2 X/2. Strategic Plan for Biodiversity 2011–2020.*

United Nations (2013). *SEEA Experimental Ecosystem Accounting.* New York: United Nations.

Walpole, M., Almond, R.E.A., Besancon, C., Butchart, S.H.M.,Campbell-Lendrum, D., Carr, G.M., Collen, B.,Collette, L.,Davidson, N.C., Dulloo, E., Fazel, A.M., Galloway, J.N., Gill, M., Goverse, T., Hockings, M., Leaman, D.J., Morgan, D.H.W., Revenga, C., Rickwood, C.J., Schutyser, F., Simons, S., Stattersfield, A.J., Tyrrell, T.D., Vie, J.C., Zimsky, M. (2009). Tracking Progress Toward the 2010 Biodiversity Target and Beyond. *Science, 325*, 1503–1504.

Wentworth Group (2014). *Blueprint for a Healthy Environment and a Productive Economy.* Sydney: Wentworth Group of Concerned Scientist.

Wilson, E.O. (2006). *Creation. An Appeal to Save Life on Earth,* New York: Norton.

WRI (2005). *The wealth of the poor. Managing ecosystems to fight poverty.* Washington D.C.: World Resources Institute.

WWF (1997). *Indigenous and traditional peoples of the world and ecoregion-based conservation: An integrated approach to conserving the world's biological and cultural diversity.* Gland, Switzerland: World Wildlife Fund International.

WWF (2014). *Living Planet Report 2014. Species and spaces, people and places.* World Wildlife Fund.

CHAPTER 4

What Are the Economic Implications of Overshoot, if Any?

By Mathis Wackernagel

INTRODUCTION

Overshoot occurs when occupants of an ecosystem take out more from the ecosystem than it can renew. Global overshoot would therefore be the situation where, on aggregate, humanity's demands exceed planetary regeneration. Global Footprint Network's *Ecological Footprint* and *biocapacity* accounts for nations are designed to measure the extent of global overshoot (Wackernagel et al. 2014). These accounts build on the recognition that life competes for biologically productive areas. Given that we live on a finite planet with a limited biologically productive surface, and given that life competes for such space, the Ecological Footprint approach asks: *How much do people demand from those surfaces (Footprint), compared to how much the planet can regenerate on those surfaces (biocapacity)?* It uses the basic scientific method of describing "what is" in a way that is testable. The assumption is that such scientific descriptions enable action, particularly if their meaning is understood and appreciated (Wackernagel 2015). So, what does it mean not to have enough biocapacity? And what is the evidence? These are the questions this chapter explores.

4.1 WHAT'S THE PROBLEM, OFFICER?

Footprint and biocapacity results have shown consistently that humanity has for some decades been operating in global overshoot. By now, these conservative estimates reckon that humanity's demands add up to at least one and a half fold what the Earth can yearly renew (Global Footprint Network 2014, WWF/Global Footprint Network/ZSL 2014). Other research confirms that humanity is operating in global overshoot, most prominently the "nine planetary boundaries" approach (Rockström et al. 2009).

Using more than what the biosphere can renew means that inevitably something has to give. For global overshoot to occur, somewhere, something is overused – whether it is depleted groundwater, declining fisheries, deforestation and reduction of forest stocks, soil erosion, biodiversity loss, or carbon accumulation in the atmosphere.

Human harvest of biomass resources has been growing and has now reached about 70 percent of planetary capacity, making up about half of humanity's Ecological Footprint (Global Footprint Network 2014).[1] Fossil fuel demand has become a dominant demand on the biosphere and has been growing even faster than the former. Furthermore, it is this growing use of fossil fuel that has

---

1   This means that the sum of all non-carbon Footprints corresponds to about 70% of the planet's biocapacity for 2010. Also the amounts of biomass use, as measured in total global hectares use of biomass based demands has increased 74% from 1961 to 2010.

enabled humanity to somewhat contain its biomass demand. There are several reasons. One is that fossil fuel can substitute traditional biomass products (e.g., synthetics instead of cotton); another reason is the ability of fossil fuel inputs to stimulate bioproductivity (fertilizers, tractors, pumping of water). So what could be the consequence if not enough fossil fuel is available? More specifically, how much could biocapacity be reduced if the growing carbon concentration in the atmosphere shifts the climate unfavorably and fossil fuel is not available to boost biocapacity?

According to IPCC-published data on carbon sequestration, there is currently enough biocapacity on the planet to sequester what is being emitted from fossil fuel burning, but then there would not be enough space to produce enough food and fiber. This underlines the constraints that are real as of today.

The physical dimensions of global overshoot seem obvious and well-documented. However – what are the economic consequences of global overshoot? Some studies predict that climate change will have significant economic impact (The Global Commission on the Economy and Climate 2014). But what about water, food, fiber, etc.? And what about not just potential impacts in the future, but measureable impacts on today's economic performance?

This question is explored in this chapter. The inquiry follows Jorgen Randers' life-long quest to reveal physical feedback of planetary constraints on socio-economic outcomes such as reduced food availability or forced population decline. But is there already evidence of global overshoot generating socio-economic impacts? And if not then why, given that global overshoot is, by any measure, so massive?

4.2   WHAT ECONOMIC IMPLICATIONS CAN WE EXPECT?

Global overshoot will probably persist for quite some time. One likely consequence would be that more and more countries would run up *biocapacity deficits* by trading more from other countries than these countries themselves are able to regenerate. Given this growing competition for biocapacity, I would expect to see economic impacts of global overshoot already today. Or I would at least expect reactions by key economic actors. But I do not see evidence of a significant enough response by such actors to address their resource situation: resource demand trends of national economies are not buckling, and the overwhelming majority of countries continue to increase their Footprint to biocapacity ratio. (This means that some are increasing their biocapacity deficit while the others are decreasing their biocapacity reserves). In other words, this evidence points towards insufficient corrective feedback by markets, technology or governance to counteract the growing or even just persisting overshoot.

Perhaps there is no need to react to overshoot since there may come a point where human activities – by means of substitution – will no longer be dependent on natural capital. However, this is not put forward by any country as their underlying assumption. Furthermore, given that every single value chain still fundamentally depends on natural capital, I hold full decoupling of economic activities from physical reality and from natural capital dependence as a rather remote possibility for many decades to come.

Given that overshoot persists and that neither markets nor technologies seem to self-correct the situation, what would be the economic consequences of translating global overshoot into specific, measurable economic outcomes?

I see primarily three potential mechanisms:

▶ *Efforts to secure everyday supplies increase for economic actors.* As physical availability of natural capital decreases, either the physical and time effort to get these materials increases per unit retrieved (i.e. the "Energy Return on Investment" decreases). In the case where it does not, the resource owner, recognizing the increased future utility of their remaining stocks, will be more reluctant to provide them to the market at today's low prices. If the market plays, both phenomena translate into higher market prices for the same goods. In other words, the resource portion of goods will overall become more expensive relative to other goods. Energy and food prices will be most strongly affected since the resource costs as a portion of their overall cost structure are particularly high. Traditional economic analysis would describe that outcome of increasing prices for the same goods as "inflation," particularly for everyday goods, even though it is not driven by excessive monetary liquidity (conventionally considered the main driver of inflation).

▶ *Value of existing assets that depend on supplies decreases.* Because large scale assets (factories, farms) often depend on physical inputs such as water, energy or fibers in order to operate, their overall utility can decrease when the access to these physical inputs becomes more difficult. For example, an airport with less traffic volume decreases in value, or a large energy-intensive house far away from urban centers decreases in value compared to the compact, efficient, well-located house as ecological constraints increase. This means that many assets could well lose market value in times of higher resource constraints. Traditional economists might describe this with "deflation."

This combination of inflation for consumer goods and deflation among asset values becomes a combined downward pressure on economies. It erodes a society's wealth at the same time as price pressure for everyday goods increases. This crunch is amplified by a third risk:

▶ *Labor loses vis-à-vis natural capital owners.* As the human population increases while natural capital assets do not, the physical scarcity balance is shifting, potentially increasing the value of natural capital, and reducing the value of labor. This may manifest as downward pressures on salaries or increased unemployment. An example may be Egypt where a growing population is meeting very limited physical resources – and youth unemployment is high (36%).[2] Some may contend that youth unemployment is even higher in Greece (54%), Spain (54%) or Portugal (38%), countries with much lower fertility rates than those of Egypt. While there are a number of possible reasons behind youth unemployment, it is still unlikely that Greece, Spain and Portugal would have less challenging youth unemployment even if they had more youth.

All three phenomena combined would suggest significant economic strain: more unemployment, less opportunities for low-income and high-income segments. Low-income segments may suffer more from everyday costs going up and the job market getting tighter, and higher-income segments may see some of their assets erode. Traditional economic analysts may underestimate the impact since, on average, inflationary effects at the consumer goods level may be compensated by deflationary effects on the asset side ("real estate drops in value, reducing housing costs – similar to the mortgage crisis of 2008 – particularly in the USA, but this reduction in housing costs might get compensated by higher food and energy costs").

Is this really the case? Does the data show that these phenomena play out?

4.3 WHAT DOES INTERNATIONAL DATA SUGGEST?

Having outlined the physical challenges of the global economy, and declared the expected mechanisms of how these physical phenomena would play out economically, what data supports or rejects this preconception?

This section does not provide a conclusive answer, but explores two countries where I would expect to see economic trends in sync with my expectations, or falsifying them.

For this purpose, I will look at national economies. National economies are not prime actors per se, but they provide a unit of analysis, typically manage

---

2   http://data.worldbank.org/indicator/SL.UEM.1524.ZS, World Bank reporting ILO statistics on "Unemployment, youth total (% of total labor force ages 15–24) (modeled ILO estimate)."

their own currency, and offer statistics that capture the entirety of eco-nomic activities within that territory. Also, the sum of all economic actors within a country face the economic challenge of how to compensate for the difference between what they use physically from nature (their Foot-print) compared to what their own ecosystems can provide (biocapacity). Of course, trade is one option to address this gap, and is an important mechanism to help take advantage of specialization. At the same time, trade is not a net-producer, only a re-distribution system. Further, the logic does not only apply to biocapacity but also to resources from the lithosphere such as fossil fuel. For instance, if a national economy is not able to extract all the consumed fossil fuels from its own territory, the country's actors, in net terms, need to purchase fossil fuel from elsewhere. This will weigh on the country's balance of payment.

For this chapter, I chose two country examples: Haiti and Egypt. Both are faced with obvious physical resource challenges. To what extent might these challenges also impact their socio-economic performance?

HAITI

Haiti is undoubtedly among the most resource-challenged countries on the planet, having lost nearly all its forests and struggling with high population densities. For 2010, their Footprint amounted to about 0.6 global hectares (gha) per person. A global hectare is the biocapacity of a world average bio-logically productive hectare, and serves as the common measurement unit for Footprint and biocapacity accounts. The Footprint of consumption of about 0.6 gha per person is just about one-fifth of the global average Footprint. But their biocapacity is even less, amounting to about 0.25 gha per person, which is just about 15% of what is available worldwide on average.

Haiti's GDP (or income base) for the same year, according to the IMF, was about 1,500 USD per person, expressed in purchasing-power-parity adjusted (or ppp) dollars. I use ppp dollars because they reflect what people supposedly are able to purchase in local markets with their income.

Figure 4.2 suggests a positive upward trend with the per capita sit-uation. As the figure shows, IMF projects for Haiti an even more rapid expansion of purchasing power per person over the next decade (from 2012 onwards).

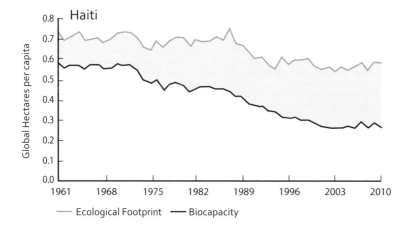

FIGURE 4.1:   *Haiti's Footprint and biocapacity in global hectares (gha) per person since 1961*. The national *Footprint* represents the biocapacity needed to provide for the average consumption of a resident. The *biocapacity* is the productive area available within a specific country. The red surface between the lines shows a growing biocapacity deficit. If the green biocapacity line is above the Footprint line, the country has a biocapacity reserve. Biocapacity deficits can be compensated by overusing *local biocapacity* (i.e. using domestic resources at a rate faster than they regenerate) or by using *biocapacity* from abroad, for instance through net-import. More country comparisons are available at www.footprintnetwork.org. Source: Global Footprint Network, National Footprint Accounts Edition 2014.

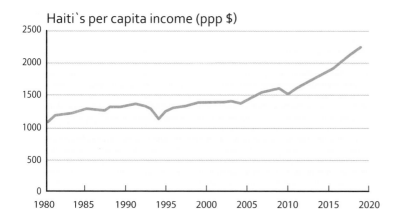

FIGURE 4.2:   This graph shows Haiti's gross domestic product based on purchasing-power-parity (PPP) per capita as reported by IMF. It is available at the IMF website. Also note that for Haiti, all numbers are IMF estimates.[1]

---

1   The data is available at http://www.imf.org/external/pubs/ft/weo/2014/02/weodata/index.aspx or directly at http://www.indexmundi.com/haiti/gdp_per_capita_(ppp).html for historical figures. For most countries IMF has historical numbers up to 2010, and the remainders are IMF estimates. For Haiti, all numbers are estimates.

EGYPT

How does the situation present itself in Egypt? Since President Mubarak's ousting in 2011, the conditions have not really recovered. For instance, international tourism has dropped significantly since 2009.[3]

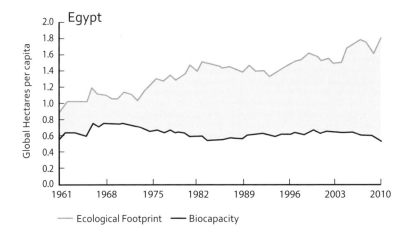

FIGURE 4.3: *Egypt's Footprint and biocapacity in global hectares (gha) per person since 1961.* The national *Footprint* represents the biocapacity needed to provide for the average consumption of a resident. It has by now grown to nearly three times the size of the domestic *biocapacity.* The red surface between the lines shows Egypt's growing biocapacity deficit. Source: Global Footprint Network, National Footprint Accounts 2014.

Egypt's Footprint, its physical demand for nature's assets, per person has continued to increase slightly, while biocapacity per person is by now only about one third of what Egypt uses. In spite of rapid population growth, biocapacity per person has remained steady since 1980, mainly due to the more aggressive use of fossil water for irrigation. (Fossil water is groundwater that is not recharging and has been sealed for thousands or even millions of years.)

---

3   http://www.nytimes.com/2014/05/09/world/middleeast/egypts-tourism-industry-grows-des-perate-amid-sustained-turmoil.html?_r=0 *The New Your Times* mentions a 30 percent fall in number of tourists. Another story in *The Guardian* claims a 95% drop in pyramid tourism revenue since 2010. Revenues from ancient Egyptian monuments such as the pyramids have fallen by 95% since Egypt's 2011 revolution, the country's antiquities minister has said. Revenues fell from 3bn Egyptian pounds (£250m) in 2010 to just 125m (£10.5m) in 2014, Mamdouh el-Damaty told al-Mehwar, a private Egyptian television channel. http://www.theguardian.com/world/2014/aug/29/egypt-tourism-revenue-falls-95-percent

At the same time, IMF reports a rapidly growing ability on the part of Egyptian residents to acquire products and services.

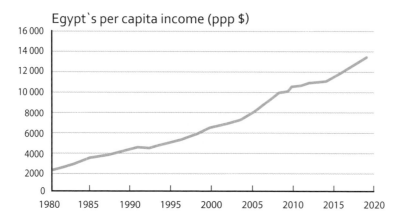

FIGURE 4.4: This graph shows Egypt's gross domestic product based on purchasing-power-parity (PPP) per capita as reported by IMF. It is available at http://www.imf.org/external/pubs/ft/weo/2014/02/weodata/index.aspx or directly at http://www.indexmundi.com/egypt/gdp_per_capita_(ppp).html for historical figures. For most countries, including Egypt, IMF has numbers up to 2010, and the remainders are IMF estimates.

INTERPRETATION OF HAITI AND EGYPT FOOTPRINT AND GDP NUMBERS

In both cases, the physical constraints do not seem to significantly impact the countries' economic expansion (even per capita) as reported by IMF. In Haiti, per capita level of economic activity is still at a low level, and expansion has been slower. Egypt, again according to IMF, seems to go through a very rapid expansion, even per capita, and in purchasing power. Also the IMF projections (2012–2020) look surprisingly rosy.

What could this apparent contradiction mean? Does the physical description of Egypt (Footprint and biocapacity, newspaper reports about increasing economic challenges in Egypt[4]) contradict GDP trends as reported by the IMF?

How reliable is officially reported GDP information? It is one of the key indicators used by economic planners. Could IMF's number conceivably be

---

4   http://www.irinnews.org/report/95262/egypt-rising-poverty-threatens-gains-in-fight-against-tb reports that increasing economic difficulties undermine efforts to contain TB. http://www.irinnews.org/report/94414/egypt-revolutionary-dreams-turn-into-economic-nightmare discusses the increasingly difficult lives for low-income Egyptians.

manipulated because GDP results have a strong influence on a country's ability to raise international debt?

The conclusion is clear: there is no clear conclusion. Current figures as reported by international institutions do not document that ecological constraints had an impact on a country's ability to expand their economic activities. But neither do economic time trends provide convincing evidence that they represent the situation of a country as reported through anecdotal evidence or by the physical constraints mapped with resource accounts like the Ecological Footprint.

Maybe it is reasonable to rely more on physical accounts, since they are not under political pressure to show particular outcomes?

Given that current numbers do not seem to produce a clear conclusion, is it still possible to show a growing economic risk emerging from increased resource overuse?

This is what the second part of the chapter outlines.

4.4   WHAT DOES THE PHYSICAL PERSPECTIVE SUGGEST?

Which countries are more likely to be robust performers in an ecologically constrained world?

TAKING THE CURRENT CONTEXT AS A GIVEN

It may by now be a safe assumption to take global resource trends towards increasing global overshoot as a given for the near future. These trends are not shifting quickly – they have large inertia. This means that ecological constraints are tightening while humanity's multiple demands on nature are increasing. The current level of humanity's material metabolism is unlikely to be sustained, unless some radical new breakthrough-technologies such as safe and economic fusion or radically cost-reduced renewable energy production become widely adopted. As a result, the safe operating space for the global economy is shrinking, and global overshoot is already the new context.

This may still mean that resource costs for high income countries will remain relatively low for some time, as current economic conditions incentivize those economies that provide many of the natural resources to liquidate them – even defying Hotelling's rule.[5] Yet, those economies receive low por-

---

5   A rational economic analyst may argue that it makes sense to sell the stock quickly if the anticipated price increase is less than future interest rates. (See Norman, this volume.) Early exploitation would therefore maximize wealth. However, it does not seem that many actors follow this rule. Particularly national governments depending on royalty income to balance their budgets will focus primarily on income generation, even if it comes at the expense of wealth.

tions of the value created in the value chain. In other words, very little of the value added flows back to the natural capital providers.

Without prices signaling increasing ecological constraints, resource costs remain relatively low, increasing the risk of future supply disruption. This has happened in the past as in the case of passenger pigeons or buffalo meat in the US, where relatively steady and low market prices for that meat did not predict resource exhaustion (Farrow 1995).

There is a significant likelihood that the physical resource constraints will worsen, and that this new context will become ever more significant. For instance, hardly any viable alternatives exist at scale to replace fossil fuels in the short to medium term. We witness continued demographic growth, with scarcely any slow down (Gerland et al. 2014), while still witnessing significant consumption deficits for billions. Overstretched ecosystems are common, with reports of soil loss, sinking groundwater levels, and declining wild vertebrate population sizes (WWF/Global Footprint Network/ZSL 2014). Furthermore, pressure from climate change may also increase. Physical systems have large inertia, leading to significant time delays. In particular, infrastructure does not shift quickly, which means current demand cannot be reduced swiftly without losing economic productivity or value.

Given the new contexts – which budget allocations and investments are likely to be the most resilient and robust performers? Or more specifically and practically, if we do not want to lose our money, on which countries should we bet?

GLOBAL COMPETITION
– FROM A FACTORY WORLD TO A GLOBAL AUCTION

To answer the question, we need first to understand how the new physical context is shifting the economic context. The essence is the following: As demand for ecological services is increasing globally, it becomes ever more relevant to identify whether a national economy (i.e. the sum of all its economic actors) is gaining or losing its economic bargaining power vis-à-vis all the other actors. This would be measured by its actors' ability to bid for resources against everyone else in the world.[6]

In a world of unlimited resources, additional demand stimulates additional supply. If more books, shirts, iPhones or potatoes are purchased, more books, shirts, iPhones and potatoes will be produced. In such a world, all that matters is the purchaser's *absolute income* – more income will give the purchaser more

---

6   Transnational companies would probably think more from the perspective of dollars, while the experience for households is better reflected by income measured in ppp.

of what she prefers. This becomes a measure of their economic bargaining power – in a demand-driven world (i.e. a world that is limited by the amount of market demand).

However, in a world of ecological constraints (energy, climate, water, food), with more and more countries running up biocapacity deficits, and depending more on biocapacity from elsewhere, the increasing demand for global biocapacity increases the competition for those services. This is turning into a world of supply limitations. In such a world, competition for ecological services could lead to additional strain, including political conflicts or military might to secure access to resources.

Let's also assume for argument's sake a most benevolent possible case. The assumption is useful because in all other cases, the effects would be even further amplified, as ecological constraints would become more disruptive. A benevolent world means that in spite of increasing physical constraints, a well established global economy would continue to operate and follow (most) WTO rules, and that financial promises and commercial transactions would continue to be honored.

Still, in such a benevolent world with transparent, equal access markets prevailing, the market dynamic of an ecologically constrained world would be shifting from a *demand-driven market*, to one that behaves more like an auction with limited goods, i.e. a *supply-limited market*. In such a world, it would be the trend in *relative income* that matters, rather than absolute income. What I mean with "relative income," and explain further below, is how much the actor earns compared to the total earnings of the other market participants.

Consider the following thought experiment. Imagine you live in a space station. Every day the space station gets a load of 50 potatoes. Every day, they get sold off to the astronauts, including you. Imagine therefore that you are sitting with the other astronauts around a table, bidding for the 50 potatoes that are supplied every day. Also imagine that every day, one more astronaut is added to the space station. Maybe you do not want to buy all the potatoes you possibly could, because you can only eat so much. But will it get easier or more difficult for you over time to bid for your share of potatoes?

Obviously, assuming this is a civilized space station, where potatoes are shared not by force but through a market mechanism (ground control is organizing an open bidding for the potatoes), your ability to get potatoes depends on your purchasing power. What matters is not only your absolute amount of money, but your relative ability to bid (i.e. your relative income) compared to everybody else on the space station.

Let's assume that you were sent to the space station by a high-income country: you are fortunate to get $100 every day while each of the other astronauts only gets $10 per day. You do not want to spend it all on food; some you will want to spend on entertainment such as iTunes, e-books and some streamed movies. Still, because of your higher income compared to the others, you will be able to get a bigger share of the potatoes on average than your co-astronauts.

Now consider this scenario. Assume that there is a salary adjustment. You get your salary doubled to $200 a day. All of the others will get a four-fold increase and receive now $40 a day each. And by now there are also more astronauts who have arrived at the space station. Also assume that the prices for iTunes, e-books and movies have stayed the same (they are beamed to the space station on demand).

What does this new situation mean for you? Even though you have in absolute terms more income at your disposal, your ability to successfully bid for potatoes has been weakened. In fact not only do you have a higher absolute income, you also still earn significantly more than the others (you earn now $160 more than the others, up from $90), and still you are only able to get fewer potatoes. *This is the auction paradox.* This paradox exemplifies the shift in importance of absolute income to relative income in a world characterized by supply constraints.

In essence, your bargaining power is linked to your income compared to the sum of all the income around the space station's bidding table. This means your bargaining power changes with your relative income compared to what everybody else around the table has available. Now consider planet earth to be such a space station.[7]

Simply put, *relative income and change in relative income,* becomes the determinant of your relative bargaining power. This means that the question on spaceship Earth becomes: how much income do I have compared to the world's total income – and how is this income changing over time compared to the income the world generates?

How is relative income changing? Figure 4.5 shows the example of Switzerland. It shows the absolute (red dotted line) and relative (blue solid line) income situation for the Swiss resident, on average. Relative to total world income, the Swiss resident has lost, on average, 50 percent of her global income share from 1980 to 2012. Yet, if we look at the Swiss person's biocapacity deficit (Figure 4.6), her average demand from the world ("the potatoes she consumes") has not diminished.

---

7   Please note that this is based on a highly benevolent world where there is full and equal market access, no bilateral treaties, no geopolitical/military power that can divert resource flows, no trade of other goods but only exchange of goods for money. Also note that without these "benevolent conditions" the competition would be even tighter and supply constraints even more disruptive for market participants.

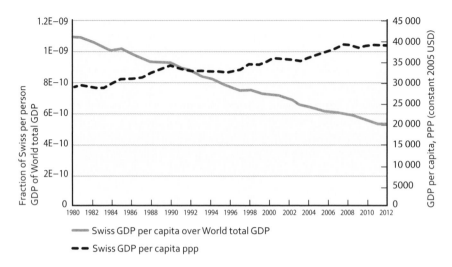

FIGURE 4.5:   Average income of Swiss residents: absolute and relative to total global income from 1980 to 2012. Data from Feenstra (2013).

I use the example of Switzerland here as representative of a high-income economy that is considered fairly robust, to emphasize that these effects do not only affect countries with challenging resources and financial situations (such as Haiti and Egypt) but have potential implications for all participants in the global economy.

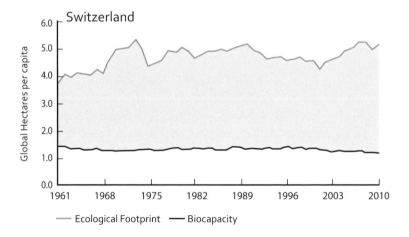

FIGURE 4.6:   The Ecological Footprint and biocapacity of Swiss residents, on average. From 1961 to today. Global Footprint Network National Footprint Accounts Edition 2014.

Figure 4.7, with the variations of the graph in Figures 4.8 and 4.9, show simplified trend lines from 1985 to 2009. These graphs put the development of
relative income in relationship with the changing biocapacity reserve (or deficit)
of countries. Countries overall are moving toward the left (larger biocapacity
deficits or less biocapacity reserve), while decreasing in relative income share.

In essence, we can observe the following. More countries want physically
more from the world. Even if global markets remain relatively open and
benevolent, this dynamic becomes more like an auction. How much does the
average person of the country earn, compared to the world total? And how is
this changing? Is your income increasing compared to world total, as your
demand for resources is increasing?

Biocapacity Deficit and Global Income Share (1985–2009)
©Global Footprint Network with data from National Footprint Account
for x-axis and WB and IMF for y-axis

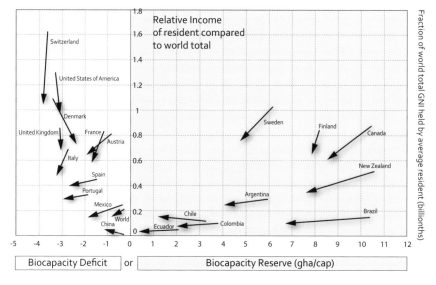

FIGURE 4.7:   The *y-axis* shows the fraction of the world's GDP (or GNI) an average resident
of a given country generates. Therefore, the world average resident's share, per definition, is at
(1/world population) or currently at about 0.14 billionths of total world GDP. The x-axis shows
to what extent a country's Footprint of consumption exceeds its own domestic biocapacity
per capita. If its Footprint is smaller, the country has a biocapacity reserve, and if it is larger,
it runs a biocapacity deficit. The arrows show trends from 1985 to 2009. For most countries,
the Footprint to biocapacity ratio has become lower, or even more negative, while relative
income has decreased for most of the countries as well. China and Chile are exceptions for
this time-period.

To keep being successful in accessing the dwindling natural capital from abroad would become costlier for economies whose relative income decreases compared to the world (as is happening for most high-income countries vis-à-vis the emerging economies). Therefore, those high-income countries (and most other ones as well) are caught in a bind: less and less relative income for their economic agents as they are competing in the purchase of ever-scarcer biocapacity.

In a world where resource access is essential for economic activities and the increased competition for resources could drive costs – even to a level where they might turn into a significant factor of economic performance – this double trap will become ever more a determinant of economic success – or failure.

Biocapacity Deficit and Global Income Share (1985–2009)

©Global Footprint Network with data from National Footprint Account for x-axis and WB and IMF for y-axis

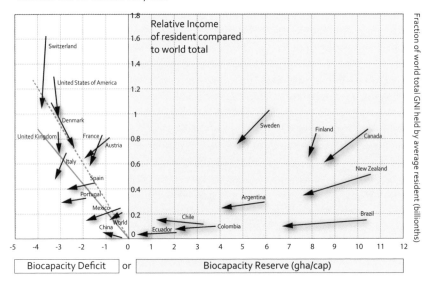

FIGURE 4.8:   The "equi-risk lines" for Denmark (dotted) and the UK (solid).

HOW TO INTERPRET THIS GRAPH,
INCLUDING THE SOLID AND DOTTED LINE?

▷ In a world of increasing resource competition, being on the biocapacity reserve side is an advantage. Globally, there are about 1.7 global hectares of biocapacity per person in 2014. Countries with relatively high standards

Biocapacity Deficit and Global Income Share (1985–2009)

©Global Footprint Network with data from National Footprint Account
for x-axis and WB and IMF for y-axis

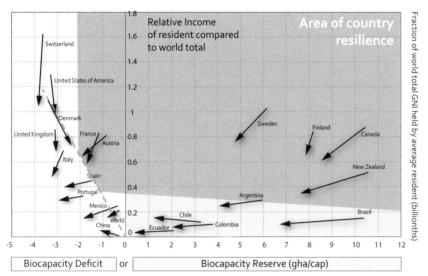

FIGURE 4.9:   The shaded area indicates where the more resilient countries are located.

of living – climate controlled houses, animal-based products in food, and good access to transportation – have typically Ecological Footprints of 4 global hectares per person and above with current technology. With better resource-efficiency, this amount could be reduced without a loss in comfort. Therefore, I consider a biocapacity of more than 4 global hectares per person to be a significant endowment, and a physical situation that should make it possible to generate relatively comfortable lives. A value of 3 global hectares per person (gha/cap) still allows for comfortable lives, if advanced technology is applied, but of course, having access to only 3 gha is more challenging than having 4 gha available.

▹ But the trend also matters. Rapidly losing one's biocapacity reserve (for instance as in the case of Ecuador, and also for Australia – even though a significant reserve still remains) is a risk. It points to future instability and the inability of the country to curb these trends.

▹ For countries with a biocapacity deficit, like the UK, it is also significant to consider what trend they are on. The UK lost some relative income, while slightly reducing its biocapacity deficit. The solid blue line shows for the UK at which angle it would have to move to maintain its ratio between biocapacity deficit and relative income. In the UK, the trend indicates that the ratio is still getting less favorable (The UK trend is steeper than the solid line). For Denmark, the ratio stays about the same (compare to the brown dotted line for Denmark), and the ratio is more favorable than for the UK. Please note though, maintaining this ratio of relative income to biocapacity reserve (rather than increasing it – or making the lines steeper) may not be sufficient to protect one's economic bargaining power. The reason is that global overshoot is still growing, increasing the competition for access to biocapacity. Maintaining the ratio is only good enough if global overshoot stayed at the same level.

▶ Also the income level matters – at lower income levels, proportionally more of the income is needed for purchasing resource inputs than at a high income. Therefore the interpretations from above are simple approximations.

WHICH COUNTRIES PROMISE TO BE MORE ROBUST PERFORMERS?

This map only evaluated the position of countries from a biocapacity resource and relative income perspective. This complements but does not replace an analysis of the financial resilience of a country. (In the financial resilience analysis, traditional factors such as the country's debt situation, productivity, inflation, unemployment or even educational levels, etc. are considered).

The shaded area approximates which countries are considered fit performers from a biophysical perspective.

From a renewable resource perspective, countries on the biocapacity reserve side are favorable. If they have sufficient resources, but are not able to generate a minimum income level, they must be suffering from a structural problem and cannot be considered robust performers either. Also, if they move rapidly towards losing their biocapacity reserve, it also shows that they are not investing in their resource resilience and that they are losing long-term potential. Since biocapacity is a decentralized asset as opposed to centralized assets such as mines of fossil fuel sources, I doubt that the "resource curse" phenomenon can be blamed for this trend.

The line is slanted slightly downward because larger biocapacity reserves offer more long-term economic opportunities, if managed well.

On the resource deficit side, a minimum level of biocapacity provides some basic resilience for a country as this level of biocapacity provides the country's economy with higher resource security. It is physically or technically difficult to produce high socio-economic outcomes on very low per capita Ecological Footprints with the current technology. 3 gha per person make it reasonably possible, but this is still far from the 5 gha per person used in Western Europe – including "resource efficient" Switzerland with hydro and nuclear powered electricity and the highest energy-efficiency standards in the building sector. As an initial approximation, it may be reasonable to cap the maximum allowable biocapacity deficit at 2 gha per person.

But there also needs to be a minimum income to biocapacity deficit ratio. We arbitrarily use France, Austria, Denmark and the US as a benchmark. Using their ratios may be too low of a ratio to truly mark safe territory.

These three lines:

▶ the 2 gha/ per person maximum deficit
▶ the income to biocapacity deficit ratio of at least Denmark or Austria, and
▶ the slightly slanting minimum Footprint line indicating minimal economic activities

define the approximate space (shaded area in Figure 4.9) containing countries that could be considered robust from a resource perspective.

4.5   WHAT ARE THE OPTIONS FOR BIOCAPACITY DEBTORS?

At least for high-income countries, several options stand out to address their biocapacity deficit risk, assuming that such deficits are indeed a risk to their economic competitiveness and economic stability *and* that this risk is being recognized as a significant threat (currently most economic strategists of national government do not seem to recognize the risk as significant). If countries like Switzerland, for example, did recognize the risk, possible options include (BakBasel and Global Footprint Network 2014):

▶ *"Retreat from the world"*: Reduce global integration as much as possible (even if it reduces standards of living) so as to avoid the negative impact of cut-throat competition over resources. Any given country might be rationally attracted to reducing its global integration. But this could be too risky since one cannot escape global markets easily or change its rules alone. Such withdrawal might not stop the global game or immunize the country against forthcoming negative impact. Withdrawing too early could indeed be very costly.
▶ *"Embrace hyper-growth"*: Accelerate your economic output in order to keep up with emerging economies. To succeed in the resource competition requires increasing the relative income of the country's residents for as long as possible. But accelerating economic expansion for any country and keeping up with emerging countries over the long haul may prove

difficult. It would have to be done using strategies that cannot be easily copied by others. Also, if the strategies would require more resources in order to succeed, the increase of economic advantages would need to be even faster.

▶ *"Hedge your bets"*: Keep maximizing the global integration benefits through a strong country brand[8] as long as it lasts, and set-up a sovereign fund as an insurance. A sovereign fund (capitalizing on the current economic advantages) needs to be large enough to allow the country to reengineer its economy when it becomes necessary. It gives the country the means to react once required. There is still risk involved since adjusting later may be cheaper (due to improved technology) or more costly (reengineering infrastructure takes time and reacts inadequately to politically volatile resource contexts). Is there enough political will to divert significant income streams into a sovereign fund?

▶ *"Reengineer extreme resource-efficiency right now"*: This means employ most efficient technologies to make the country far less dependent on foreign resources, for instance by aggressively applying circular economy principles, amplifying production efficiency and investing in natural capital. Such an approach would also ensure a soft landing out of today's form of globalization should that become necessary. It is a move from Paul Samuelson's "more butter, less guns" to "less butter, more post-oil infrastructure." The additional challenge is to get there without losing labor productivity. The latter is a particular concern since much of the labor productivity increase has been gained by having access to cheap resources and energy. Without such a pre-condition could high wages be sustained?

Another variant of this strategy may be to invest heavily in the resource efficiency of value chains leading into the country.

▶ *"Forge privileged resource relationships"*: One way of securing the country's supply may be to develop long-term bilateral resource contracts with biocapacity rich nations. Enabling this would require significant additional intervention by the government (since hitherto most resources are traded privately and not via government-sponsored channels). This strategy will only work if a) these countries respond to one's special interest, contracts can be maintained without political vassalage and inconsistency with other international obligations, and only if transport lines are secured, and b) the ensuing special relationship with the supplier country is found

---

8   Currently, a large portion of the value-added of a supply chain goes to the brand-holder of the final product. This underlines the significance of maintaining a strong brand. But it is not clear whether brand-advantages can be maintained over the long-run as ecological constraints tighten.

acceptable. Negotiating such long-term contracts may be politically challenging, including on the domestic front, and would require significant government investments.

Neither of the above options can be easily embraced as the obvious fix. Is there a sixth option, not yet envisaged? Or is this sixth option an optimal blend of the five previous ones: "All of the above"? And if yes, again, what are the affordable pathways to a resilient, overall resource security? In particular, what is the optimal biocapacity deficit the country wants to achieve?

CONCLUSION

In conclusion, while current impacts of ecological constraints do not show up clearly in the currently available statistics, there is an emerging clear systemic risk from the growing biocapacity deficits combined with the vanishing relative incomes.

There are also ways for countries to react to those risks. Whether the therapy, though, proves more painful than the disease is something each country needs to explore for itself.

BIBLIOGRAPHY

BakBasel and Global Footprint Network (2014). The Significance of Global Resource Availability to Swiss Competitiveness, *ARE*, Bern downloadable from http://www.are.admin.ch/dienstleistungen/04135/05243/index.html?lang=de

Farrow, Scott REF Ecol Econ study (1995). Extinction and market forces: two case studies, *Ecological Economics*, *13*(2), 115–123.

Feenstra, R.C., Inklaar, R. and Marcel P. Timmer, (2013). *The Next Generation of the Penn World Table*, available for download at www.ggdc.net/pwt

Gerland, P., Raftery, A.E., Šev íková, H., Li, N., Gu, D., Spoorenberg, T., Alkema, L., Fosdick, B.K., Chunn, J., Lalic, N., Bay, G., Buettner, T., Heilig, G.K. and Wilmoth, J. (2014). World population stabilization unlikely this century, *Science*. Oct 10, 2014; *346*(6206), 234–237.

Global Footprint Network (2014). *National Footprint Accounts*, Edition 2014. www.footprintnetwork.org

Rockström, J., Steffen, W., Noone, K., Persson, Å., Chapin, F.S., Lambin, E.F., … Foley, J.A. (2009). A safe operating space for humanity. *Nature*, *461*(7263), 472–475. doi:10.1038/461472a

The Global Commission on the Economy and Climate (2014). Stern et al. 2014 Better Growth, Better Climate: The New Climate Economy Report http://newclimateeconomy.report /).

Wackernagel, M. (2015). *Mathis' Advice to Science-Based Mission-Driven Organizations Concerned About the Collision Course between the Human Economy and Planetary Constraints.* Global Footprint Network. Internal Communication. Available from the author.

Wackernagel, M., Cranston, G., Morales, J.C. and Galli, A. (2014). Chapter 24: Ecological Footprint Accounts: From Research Question to Application, in: G.Atkinson, S. Dietz, E. Neumayer and M. Agarwala (Eds.), (2014), *Handbook of Sustainable Development: second revised edition.* Cheltenham, UK: Edward Elgar Publishing.

WWF, Global Footprint Network and Zoological Society of London (2014). *Living Planet Report 2014.* WWF International, Gland.WWF/ Global Footprint Network/ZSL 2014. Living Planet Report.

Part II Climate Policy Research

The second part looks into social science research on climate change policy. In Chapter 5 Per Espen Stoknes reviews the recent evidence on subjective well-being to launch the idea of "A Happy Climate? New Stories for Climate Communication." An alignment between climate and well-being is needed in policy narratives. In Chapter 6, Caroline Ditlev-Simonsen explores whether the public in general can be seen as science based activists by conducting a Norwegian empirical study on the gap between attitude and behavior in environmental protection. She shows that people are open to intervention by the authorities to facilitate sustainable development. People as a rule however lack the conviction and commitment required to act independently, preferring instead that the public authorities enforce measures by means of regulation.

The next two chapters shift from a bottom-up social psychology approach to top-down political science. The topic is how typically short-term focused democracies deal with long-term climate change. The authors look closely at the role and policies of the EU in Climate Policy because of the leading global role this organization has aimed for and to a certain extent succeeded in assuming. Chapter 7 by Nick Sitter focuses on the fascinating issue of the implication of EU Energy Policy for the climate, showing how climate change policies to a large extent in the EU have become integrated with overall energy policy. In Chapter 8, Kjell A. Eliassen, Marit Sjøvaag-Marino and Pavlina Peneva study the more detailed inner dynamics of how the 2008 "20 20 20" energy and climate package was agreed upon by the EU. The chapter argues that EU climate policy and implementation in the member states shows how we can exploit elements of the lack of perfection in democracies for implementing radical and unpopular climate policies.

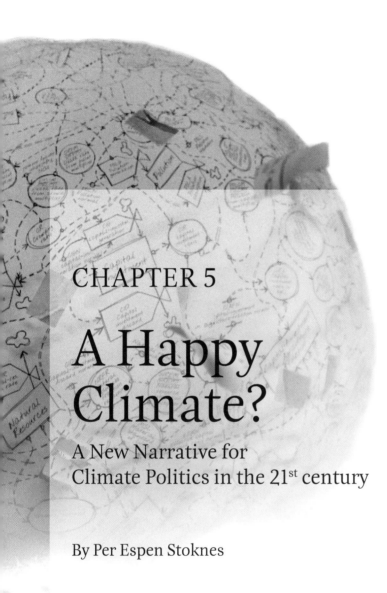

CHAPTER 5

A Happy Climate?

A New Narrative for
Climate Politics in the 21ˢᵗ century

By Per Espen Stoknes

INTRODUCTION

The latest quip from many USA republicans, when asked about their view on climate, is to say: "I'm not a scientist". This does not hold them back from expressing opinions on Ebola – without being epidemiologists, or on evolution – without being biologists, or on roads – without being engineers or city planners. But in terms of climate, they refuse to enter into the issue at all, since they are "not scientists". Thus, by distancing themselves from climate science, the case for ambitious policy, they hope, is weakened. Aversion to climate science is more often due to the policy *implications* of climate science, rather than to the climate science itself (Campbell and Kay 2014). Policies such as raising carbon taxes and stricter regulations, result in cultural cognition and ideological denial among those who prioritize stronger economic growth and unregulated markets, particularly in many English-speaking countries such as the US, UK, Canada and Australia (Kahan, Jenkins-Smith and Braman 2011; Lewandowsky, Oberauer, and Gignac 2013; Mooney 2014; Stoknes 2014).

This seems to be the polar opposite of science based activism; a claim to ignorance as a basis for policy advocacy. This reflects tactics to display an image as being practical, economy-friendly, people-caring politicians as opposed to abstruse, nerdy and unpractical scientists hell-bent on big government and carbon taxes.

This chapter reviews the growth of a new scientific narrative inside which such science-avoidance behaviours become unnecessary, since the narrative reframes and aligns the long-term objectives of society and climate.

5.1 THE OLD DEBATE ON GROWTH

The publication of *Limits to Growth* in 1972, succeeded in igniting the public debate about the (im)possibility of infinite economic growth on a finite planet. The fronts became quickly polarized into those that remained within the economic growth world-view and those that took the more physical whole-system view (see Hernes and Norman, in this Festschrift). Thus, some believed in the endless growth story, while others believed in the overshoot & collapse-story. Unfortunately, the debate quickly turned unproductive, with many labelling *Limits to Growth* as a doomsday study promoting unrealistic scenarios.

In hindsight, however, it is easier to see that the debate did not have the necessary distinction and clarity in terms of fundamental concepts and metrics. Economics measures growth in monetary terms, while the *Limits to Growth* operated with physical-based variables. There was thus a muddling – at least in the minds of the audience for the debate – of economic growth (USD) with growth in material throughput (tons). This isn't strange since in the

1900's the two were highly correlated. The two fronts became entrenched; one "believing" in endless economic growth as price signals would facilitate extensive capital substitutions: scarce resources beget higher prices which rapidly lead to increasing supply. The other camp was "believing" that physical limitations and entropy sooner or later would prevail over price mechanisms (in particular due to delays in the socio-economic system). This was pitting one science, economics against ecology and physics on the other, one worldview against another.

Since then, during the 1990s and 2000s, improved methods of accounting for the physical resources of the economy have emerged. Examples include the Ecological Footprint (Wackernagel et al. 1999) and Physical Flow Accounting in the UN system of SEEA (UNStats 2012). Further, a clearer understanding has been established of the applications and limitations of the conventional economic growth measure; the GDP (Fleurbaey 2009; Stiglitz, Sen, and Fitoussi 2009).

These developments make a conceptual resolution of the "old growth debate" possible, an end to the stalemate of the impossibility of infinite economic growth on a finite planet. Now, the previously monolithic concept of "growth" can be decomposed into "growth in economic value added in monetary terms" on one side and "growth of economy-wide physical flows" on the other. Or more simply: value added relative to resource flow. What (almost) everybody on both sides wants is *more value added* from *less resource flows* (until they fit comfortably within the long-term carrying capacity of physical planetary limits). That solves the conceptual conundrum of the old growth debate. It shifts the discussion onward to the feasibility of a strong decoupling in which growing value added can happen with lower physical flows than at present. Put differently: How much must resource efficiency grow relative to the value added, in order to secure lower resource flows while ensuring growth of the economic value?

5.2   SOCIAL WELFARE: EVER MORE UTILITY OR MORE WELL-BEING?

Endlessly increasing people's incomes and overall value added is in itself, however, not a good overall societal purpose. Yet another step is therefore necessary. And this step starts from the recognition of what many economists repeatedly remind the world of: that GDP (or value added) is not a welfare indicator. GDP per capita is not even a good approximation of citizens' well-being. It simply measures the level of economic activity, covering both "goods" and "bads". For instance, if inequality increases quicker than average per capita

GDP, most people can be worse off even though average income is increasing.[1]
Similarly, if city air pollution is severely deteriorating while average incomes
are rising, overall most people can be worse off. Still, GDP per capita has often
been used as an approximation of social welfare and well-being despite of its
many shortcomings in this regard. GDP is thus not wrong as an indicator of
economic activity as such, but often wrongly used.

The concept of *utility* is the economic basis for the conventional economic
approach to welfare. Utility is revealed by what people actually prefer to buy.
These "revealed preferences", as they were named by Paul Samuelson, show
up in people's willingness to pay:

> {*In economics,*} *utility is taken to be correlative to Desire or Want. It has been already*
> *argued that desires cannot be measured directly, but only indirectly, by the outward*
> *phenomena to which they give rise: and that in those cases with which economics is*
> *chiefly concerned the measure is found in the price which a person is willing to pay*
> *for the fulfilment or satisfaction of his desire. (Marshall 1920, p. 78)*

A very short recapitulation of the history of the utility concept can be useful:
Long before Samuelson's definition, economists were actually interested in
measuring well-being directly. In the 1700s and 1800s, the "dismal science" of
economics was actually concerned with happiness. Inspired by thinkers from
Jeremy Bentham to John Stuart Mill, many would see pursuing happiness as
the ultimate goal of economics. The first, more popular conception of utility, from Bentham through Frances Edgeworth to Alfred Marshall was that
utility could be seen as the net sum of a continuous flow of varying amounts
of pleasure and pain.

But it was not easy to find good ways to measure and weigh pleasure
and pain. In the absence of a "hedonistic happiness meter" which among
others Edgeworth tried to make, another, more generative metaphor came
up: "Welfare is to meet people's desires." And what actually satisfies people's
desires can be seen as displayed in their purchases: in the situations of choice
with limited resources (money, time), their revealed preferences are good
indicators of which goods and services are most likely to meet their desires.

The development of the economic utility theory after Bentham and Mill's
original Utilitarianism, has undergone two defining periods, claim economists Cooter and Rappoport (1984): First came the "marginalist revolution"
from 1870 onwards, and then there was the "ordinalist revolution" of the

---

1   This has been the case in the USA in recent decades since 1980.

1930s. The first turn-around established a central place for utility theory in economics. The second further restricted the kind of utility that was acceptable to work with in economics. Ordinal means the relative sorting of preferences only "inside" each individual, not between two people. Cardinal means that utility can be measured in units that can be compared between different individuals. (Ordinal is thus near "internal" within a given set of preferences, while cardinal is near "interpersonal".) The question of whether a hungry man gets more utility from a meal than the bored man from a theatre ticket for the same price, makes no sense within the ordinalist view. If that is each person's choice, then both have received maximum utility. The "ordinal revolution" thus entailed a rejection of *cardinal* notions of utility.

Why is this important today? This ordinal thinking, which Samuelson helped to spread and solidify with his elegant mathematics, became the cornerstone of neoclassical welfare economics. "Revealed preference" became the ultimate criterion for judging whether or not what makes a person better off is what she consumes, in a choice in which she is well informed. This removed the economic metrics of well-being far from direct human experienced utility. After that turn-around, social well-being could be calculated – economists thought – by consumption levels: the more preferences satisfied, the higher economic utility. And then the total of consumption within the economy – as defined by Gross Domestic Product – was correspondingly quickly seized upon as a convenient measure of national achievement and progress.

Within this framework, climate policy has always been politically evaluated according to its impacts on social welfare. This meant any mitigation should start – if at all – with the most cost-effective measures that reduce people's utility the least (see Marit Sjovaag, this volume).

Another unintended consequence of this ordinal revolution in welfare economics is that social well-being can only be increased by also increasing utility through higher consumption. Further, since increased consumption is usually associated with increased use of materials and fossil fuel energy, further improving social well-being has tended to come at the expense of more climate emissions. In the 20th century there was a rock solid correlation between GDP growth and emissions growth. But at least since 2000 it has become increasingly apparent that further trashing the planet's atmosphere to further raise people's consumption levels *when that does not even increase people's well-being*, is clearly not the brightest idea that humanity is capable of. If society's goal is to increase people's measured well-being, then growth in utility and value-added in itself will not do the job.

The last twenty-five years have seen the emergence of a whole interdisciplinary science of how to measure happiness and subjective well-being. According to the literature database Econlit, there was for example, only four economic articles that made analyses based on measurements of happiness and life-satisfaction data for the period 1991 to 1995, while there were over 100 articles published in the same period ten years later, i.e. 2001 to 2005 (Kahneman and Krueger 2006). In psychology, Ed Diener et al. could already by 1999 review more than 300 articles in psychological journals over the preceding three decades scrutinizing "subjective well-being" data (Diener, Suh, Lucas, and Smith 1999). An increasing number of methods have been established to define and quantify what is meant by terms such as quality of life, happiness, life satisfaction and subjective well-being. There are now standard measuring tools and even professorships in happiness at leading international universities. The field has its own international journals such as *Journal of Happiness Studies* founded in 2000.

It is not the aim of this chapter to provide an overview of this burgeoning literature. This has been done elsewhere (Huppert, Baylis, and Keverne 2005; Kahneman, Diener, Schwarz, and Russell Sage Foundation 2003; Kahneman and Krueger 2006; Krueger and Stone 2014; Layard 2011; Stiglitz et al. 2009). The point here is to underline that the rapidly maturing discipline of the science of well-being is being positioned to clearly contribute to the policy and climate economics toolbox.

SUMMARY OF MAIN FINDINGS

Researchers have looked at the correlation between overall life-satisfaction and happiness with other behaviour and characteristics of well-being. They have found, according to Daniel Kahneman[2] that self-reported happiness is highly correlated with characteristics and types of behaviour such as:

- *Smiling frequency*
- *Smiling with the eyes ("unfakeable smile")*
- *Ratings of one's happiness made by friends*
- *Frequent verbal expressions of positive emotions*
- *Sociability and extraversion*
- *Sleep quality*
- *Happiness of close relatives*

---

2   (Kahneman and Krueger, 2006) Who in turn refer to (Diener and Suh, 1999; Frey and Stutzer, 2002; Layard, 2005).

▶ *Self-reported health*
▶ *High income and high income rank in a reference group*
▶ *Active involvement in religion*
▶ *Recent positive changes of circumstances (increased income, marriage)*

There has been particular interest in results that compare national data on well-being to growth in average income per capita. One main finding is that well-being increases along with income up to a certain point, and then flattens off to no or very little gain in happiness for each extra dollar of consumption. This has been named the Easterlin paradox: that money does not – in richer countries – buy more happiness over time (Easterlin et al. 2010; Easterlin et al. 2012; Layard 2011; Stevenson and Wolfers 2008).

Finally, the happiness research has established clear evidence that low unemployment and a strong social safety net increases happiness (Richard A. Easterlin 2013).

What emerges from the abundant scientific literature is that there are valid and reliable methods for measuring well-being both in and among individuals. Thus the science of well-being is firmly established, if not by any means complete. The traditional economic and political focus on ordinal utility – as displayed by revealed preferences, aka consumption – can now be complemented by direct measures of well-being. This makes for a kind of return "to the roots" of economics from before the ordinal revolution. The question now turns to how – and if – this science can be applied to policy development. If so, this can open up to a new type of politics based on a new scientific approach, purpose and narrative: setting an increasing well-being as the ultimate vision rather than just growing utility. A series of approaches and practical ideas have emerged so far:

5.3 THE "HAPPY CLIMATE" APPROACH

Is it possible to improve and verify well-being while at the same time lowering ecological footprints back toward planetary boundaries? Can we have higher well-being with lower resource-flows? This is the emerging question that is attracting researchers, NGOs and activists to study and actively promote it as a real alternative. Advances in the measurement of well-being mean that we can now reclaim the original purpose of economics and national accounts as originally conceived. During most of the 20th century, countries competed internationally to achieve a type of economic growth based on industrial production and consumption. But that also increased climate emissions through a tight coupling of economic growth with the use of fossil energy and thus greenhouse gas emissions. The increasing consumption of fossil fuels has been legitimized by pointing to jobs and higher living standards. The linking of happiness to climate implies that we need to shift away from primarily increasing consumption to increasing experienced well-being. What we want is a society with higher direct well-being with a benign climate.

In this climate-narrative, the emphasis is on moving toward measures which capture and promote the real wealth of people's lived experience, rather than emission-generating consumption that does not further increase well-being. The concept of well-being has two main legs: feeling good (time spent in positive affect) and functioning well (capabilities). Feelings of happiness, contentment, enjoyment, curiosity and engagement are characteristic of someone who has a positive experience of their life. "Equally important for well-being is the functioning in the world. Experiencing positive relationships, having some control over one's life and having a sense of purpose are all important attributes of well-being" (Aked, Marks, Cordon, and Thompson 2008; Huppert 2009).

Before Adam Smith, what mattered most to sovereigns was holding as much land, gold and serfs or bondsmen as possible. Smith championed the wealth of nations through markets, trade and industry. Post-war rich countries competed on GDP growth driven by production for consumption; UK was superseded by the US, which again is being superseded by China, with its huge labour forces. What drives GDP growth has been thoroughly studied (World Bank and Commission on Growth and Development 2008), and first is labour & labour productivity. If you want a big GDP, the thing to do is to increase the labour force, working hours and labour productivity (more machines, energy and new technology). Within this purpose and framework, current policies of rapidly exploiting natural resources for economic growth make a lot of sense.

But now it's human well-being and happiness that we're really after, not just a large and growing GDP. This is an updated version of Jeremy Bentham's concept of maximizing the most happiness for the most people. If so, welfare policies could learn from psychological sciences about what actually improves our lives, and which doesn't have to cost the earth. The main factors are social relations and self-respect from work well done. It also helps if you can feel you have *more than* others. This relative status is much more important than the absolute consumption levels.

Within this approach a set of new metrics and instruments become prominent in climate policies. I will here briefly review four examples of these: a) The Happy planet index, b) improving human well-being directly c) labour reform, d) investment in social and natural capital.

THE HAPPY PLANET INDEX

The UK-based New Economics Foundation has developed an index labelled the "Happy Planet Index", or the HPI to measure the progress of countries. They state that:

The Happy Planet Index is a new measure of progress that focuses on what matters: sustainable well-being for all. It tells us how well nations are doing in terms of supporting their inhabitants to live good lives now, while ensuring that others can do the same in the future.

In a time of uncertainty, the Index provides a clear compass pointing nations in the direction they need to travel, and helping groups around the world to advocate for a vision of progress that is truly about people's lives.

Since it encompasses both human well-being and ecological footprint, it can be called "sustainable well-being". The formula used to calculate it is:

Happy Planet Index ≈ (Experienced well-being x Life expectancy) / Ecological Footprint.

It first calculates the number of *happy life years*, through multiplying well-being by average life length. Then by dividing by ecological footprint, it gives an indication of how effectively society utilizes the flow of natural resources to enhance a good life. Countries that do well on this index have overall satisfied people with long lives, combined with low use of natural resources. In a nutshell, it shows the ecological efficiency of supporting human well-being.

Many leading countries on the HPI come from Latin America and the Caribbean, particularly since these countries have lower resource use, such as Costa Rica, El Salvador, Vietnam and Colombia. Costa Rica comes out as number one due to its very high life expectancy which is second highest in the Americas: it has experienced well-being higher than many richer nations and a per capita footprint just one third of the US. The Happy Planet Index has attracted a lot of criticism since it reverses or shakes up the standard order of what is seen as the most highly developed nations, usually measured by GDP income/capita. Countries such as Bangladesh, Colombia, Albania and Pakistan come out higher on the ranking than Norway, Sweden, the United Kingdom and Switzerland. The US ends up in 108th place.

To "real economists", such as the Adam Smith Institute, this qualifies as pure "madness":

It is difficult to take an index seriously when it places Iraq (36th) and Albania (18th) ahead of Iceland (88th) and Australia (76th). It is not just that the list contains some strange anomalies, rather that it defies common sense from start to finish.[3]

The HPI ranking comes as a consequence of taking the ecological limitations of this planet seriously; this implies that a high usage of resources per person actually takes its toll on the efficiency of converting natural resources into happy life years. It is an index that gives *planet* in Happy Planet Index a high weighing. Within a short-term, utility-centric worldview, this may not make much sense. However, in a long-term, eco-centric worldview like that found in the *Limits to Growth,* it may be seen as striking to the core of what really matters: the flourishing of human societies within long-term planetary boundaries.

HOW TO IMPROVE WELL-BEING DIRECTLY
THROUGH SOCIAL RELATIONS AND MEANINGFUL WORK

Building on the work of psychologists Huppert and Luybomirsky among others (Huppert 2009; Huppert et al. 2005; Lyubomirsky, Sheldon, and Schkade 2005) Aked et al. (2008) has worked out a set of five evidence-based measures that increase the individual's well-being directly. The idea was to make something like the "5 fruits or vegetables a day" policy advice. They summarized the best empirical advice in these five key areas: social relationships, physical activity, awareness, learning, and giving.

Briefly summarized:[4] 1 – *Connect:* Social relationships are critical to our well-being. Survey research has found that well-being is increased by life goals associated with family, friends, social and political life and decreased by goals associated with career success and material gains. Governments can shape policies and cities in ways that encourage citizens to spend more time – on average – with families and friends and less time in the workplace.

2 – *Be active:* Exercise has been shown to improve mood and has been used successfully to lower rates of depression and anxiety. Through urban design and transport policy, governments influence the way we navigate through our neighbourhoods and towns. To improve our well-being, policies could support more green space to encourage exercise and play and prioritize cycling and walking over car use.

---

3   http://www.adamsmith.org/research/think-pieces/shiny-happy-people-the-madness-of-the-happy-planet-index/

4   The following list contains lightly edited excerpts from (Aked, Marks, Cordon, and Thompson, 2008)

3 – *Take notice:* Research has shown that practising awareness of sensations, thoughts and feelings can improve both the knowledge we have about ourselves and our well-being for several years. Policy that incorporates emotional awareness training and media education into a universal education provision may better equip individuals to navigate their way through the overload of the information super-highway.

4 – *Keep learning:* Learning encourages social interaction and increases self-esteem and feelings of competency. Behaviour directed by personal goals to achieve something new has been shown to increase reported life satisfaction. Policies that encourage continued learning, in all ages, will enable individuals to develop new skills, strengthen social networks and feel more able to deal with life's challenges.

5 – *Give:* Studies in neuroscience have shown that cooperative behaviour activates reward areas of the brain, suggesting we are hardwired to enjoy helping one another. Individuals actively engaged in their communities report higher well-being and have knock-on effects for others. Policies that provide accessible, enjoyable and rewarding ways of participation and exchange will enable more individuals to take part in social and political life.

Aked (2008) concludes that:

> [T]he challenge for governments will be to create the conditions within society that enable individuals to incorporate these … positive activities more consistently into their daily lives. By measuring the impact of their decisions on the components of personal and social well-being, government can provide a great boost to efforts to shape the policy cycle to one which is explicitly well-being promoting.

MORE VACATIONS – LESS WORKING HOURS

It is well known from research that it is *self-respect from work well done*, not just a lot of work hours, which strongly contributes to happiness (Layard 2011). The happiness received from feedback from others on a job well done, does not suffer habituation to the extent that consumption of things does. Thus unemployment is often a disaster for self-respect, since it makes people feel redundant and worthless. When people become unemployed, their happiness falls – not just due to the income loss – but even more because the loss itself makes them seem unworthy in the eyes of others. Our sense of purpose and self-respect is critical to well-being.

Worldwide both populations and labour productivity are increasing: there are more hands than are needed as more and more work is automated. Unemployment also increases income inequalities, which is a major cause of unhappiness. Further, when there is over-production of non-durable consumer goods, the climate suffers. The natural policy consequences are to reduce the number of working hours that each person performs per week and year. This means sharing the jobs with more people.

Many economists once believed that as technology improved and boosted workers' productivity, people would choose to bank these benefits by working fewer hours and enjoying more leisure. Instead, working hours have become longer in many countries. The US has a 47-hour (!) working week.[5] The UK has the longest working week of any major European economy. Some suggest cutting it to 20 hours or so per week. "The civilised answer should be work-sharing. The government should legislate a maximum working week," says Robert Skidelsky, a Keynesian economist (Stewart 2012).

Another example for employment policy that actively promotes flexible working and reduces the burdens of commuting, alongside policies aimed at strengthening local involvement, would enable people to spend more time at home and in their communities to build supportive and lasting relationships.

Shifting the work-life balance to less salary work and more social life is good both for happiness (Layard 2011) and for climate since production & consumption then increases less.

INVESTING IN SOCIAL CAPITAL AND NATURAL CAPITAL

It may not make immediate sense why investing in social capital and natural capital would be effective policies to create a happier climate.

The first argument is that sociologists, economists and psychologists have all documented that a society with well-working institutions and high level of trust, is a happier society (Helliwell 2006; Helliwell and Huang 2008; Helliwell and Putnam 2005; Putnam 2007). These are key elements of "social capital": the internal structure in societies of hospitals, governmental agencies, police, courts, unemployment centres, churches, festivals, municipalities, neighbourhood associations, sports clubs, scouts, youth clubs, etc. What they do is strengthen the bonds and trust that secure a smoothly functioning society. Without it there is more crime, corruption and higher economic costs. It has been shown to correlate highly with happiness and well-being (Leung, Kier, Fung, Fung, and Sproule 2013; Rodríguez-Pose and von Berlepsch 2014). If it's happiness we want, make sure the institutions of the public sector and the third sector are well-functioning.

Newer happiness research has also found support for the claim that the access to natural capital – and the quality of nature accessible – clearly influences hap-

---

5   Gallup, 2014, 29. Aug: http://www.gallup.com/poll/175286/hour-workweek-actual-ly-longer-seven-hours.aspx

piness (Engelbrecht 2009; MacKerron 2011; MacKerron and Mourato 2013). Kahneman and Sugden (2005) argue for using subjective well-being measures also in environmental economics. Several studies are strengthening the argument that evaluations of natural capital and social welfare should incorporate subjective well-being measures. They increasingly seem to be important in the new welfare economics that includes climate considerations.

The happiness of people and the happiness of climate go together.

5.4   THE 21ST CENTURY: AN AGE OF HAPPINESS POLITICS?

According to the research briefly introduced above, it may seem that happiness and well-being are making strong headway into the policy development sphere. And quickly. A giant step was taken in 2009, with the Sarkozy Commission that recommended adding SWB measures as supplements to existing indicators of societal progress. Several countries in the EU, like the UK, France and including the EuroStat itself, have started implementing these measures through their surveys compiled by the statistical offices. The breadth, depth and applications of well-being data increase yearly.

Yet. Objections are sure to remain strong from market fundamentalists. Views ideologically anchored in neoclassic economics are driving a current backlash against well-being or "happiness politics". The UK conservative Institute of Economic Affairs recently published a report they named *Quack Policy*, in which Jamie Whyte attacks any use of happiness politics as "paternalist":

> *Paternalist policies promoted by experts and politicians show contempt for the actual preferences of the general public. People are forced to live according to values that they reject. For example, supporters of 'happiness policy' believe the state should coerce people to act against their preferences in ways that policymakers think will increase their well-being. (Whyte 2013, p. 13)*

The Adam Smith Institute rips the Happy Planet index apart:

> *[The proponents of a Happy Planet index,] not only doggedly insist that a lot of failed states are the new Jerusalem, but seriously suggest that we become more like them. Those who fear that far-left environmentalists use climate change as an excuse to send us back to the dark ages will find much to encourage their beliefs here. (Snowdon 2012)*

Particularly in the US, NGOs that work for strengthened climate and biodiversity are met by harsh ideological counterattacks, such as from hard-hitting

Berman & Company (Lipton, 2014). They want to portray big greens in a negative light:

> *The Sierra Club has become an anti-growth, anti-technology, anti-energy group that puts its utopian environmentalist vision before the well-being of humans – all while receiving funding from a suspicious web of donors. (Berman 2014)*

Thus there are many writers, usually with libertarian leaning, who are quite averse to the Easterlin conclusion, thinking it will lead us to adopt Luddite policies because growth would not matter in such a world (Falkenstein 2013).

Accusations of paternalism, Luddism, Malthusianism, anti-growth, anti-human, anti-freedom are sure to keep surfacing. And to a certain extent it's pertinent. Because, environmentalists have – over the last decades – drawn attention with messages containing an endless series of no's and don'ts and demands for cuts and stops. These activists – with the best of intentions, science based or not – have been perceived by the mainstream as prigs, kill-joys and party-poopers. Their message has been mainly negative, very critical of markets, growth and consumption. But with a new focus on human well-being coming into the mainstream, the fronts can potentially be reconciled, at least in the view of the voting public if not among "die-hards" on both sides.

The common way to counter all these is to develop, expand and clarify the narrative that human well-being and a better climate are two sides of the same coin. What we want is better lives *and* better climate. Caring for the climate is actually caring for people and human progress, too. In a severely disrupted climate, there will also be much less personal choice and freedom. Ambitious climate policies should be argued on the grounds that they serve personal freedom and human happiness (both in the short and the long run).

What is needed is to break the strong *mental model* from the 20th century, in which well-being is linked with increasing (material) consumption. According to psychological science, this link is very weak or non-existing in rich countries (Easterlin 2013; Layard 2011). Yet, in conventional economic theory – and in economic mental models inside people's minds, it is still strongly embedded. Only within this mental model, is decoupling material consumption growth from well-being equated with going backwards, becoming Luddites, returning to the Middle Ages. The narrative of pursuing both human well-being and climate, a "happy climate", avoids these pitfalls: By shifting the priorities and metrics to human well-being before consumption, to investment in happiness rather than material throughput, we get a better framework for aligning our societies to the planetary limits in the 21st century.

CONCLUSION

One can't win the public majority's support by arguing for environment, climate or the planet. When the public prioritizes among political issues today, what comes out on top is economy, well-being of people, good jobs, health, education and community (Pew Research Center 2012). By linking climate to well-being, health and radical resource-efficiency, the climate issue is closely integrated with these top political issues. This opens to opportunities of double-wins, which gives meaning to well-being-enhancing measures such as fewer working hours, better health from clean air and land, and jobs, products and processes that waste less.

Climate and environmental activists have been very good at saying what they are *against*: cutting emissions, less coal, less fracking, stopping population growth, cutting conspicuous consumption, stopping deforestation. Scientists and academics are good at analysis, criticism and debunking everything that seems "wrong". But they rarely enter into the area of synthesis and vision. But for science based activism it's time now to make a big swerve toward *focusing on what we want*, what we're for and how to get there. The "happy climate" framework fits this need well: We want to move quickly forwards to a society with more human well-being *and* a benign climate. It is difficult to disagree with that vision. Or what?

BIBLIOGRAPHY

Aked, J., Marks, N., Cordon, C. and Thompson, S. (2008). *Five ways to wellbeing.* New Economics Foundation. Retrieved from http://www.neweconomics.org/publications/entry/five-ways-to-well-being-the-evidence

Berman, R. (2014, November 15). *Sierra Club – BigGreenRadicals entry.* Retrieved November 15, 2014, from http://www.biggreenradicals.com/group/sierra-club/

Campbell, T.H. and Kay, A.C. (2014). Solution aversion: On the relation between ideology and motivated disbelief. *Journal of Personality and Social Psychology, 107*(5), 809–824. doi:10.1037/a0037963

Cooter, R. and Rappoport, P. (1984). Were the Ordinalists Wrong about Welfare Economics? *Journal of Economic Literature, 22*, 507–530.

Diener, E., Suh, E.M., Lucas, R.E. and Smith, H.L. (1999). Subjective well-being: Three decades of progress. *Psychological Bulletin, 125*(2), 276–302. doi:10.1037/0033-2909.125.2.276

Easterlin, R.A. (2013). Happiness, Growth, and Public Policy. *Economic Inquiry, 51*(1), 1–15.

Easterlin, R.A., McVey, L.A., Switek, M., Sawangfa, O. and Zweig, J.S. (2010). The happiness-income paradox revisited. *Proceedings of the National Academy of Sciences*, *107*(52), 22463–22468. doi:10.1073/pnas.1015962107

Easterlin, R.A., Morgan, R., Switek, M. and Wang, F. (2012). China's life satisfaction, 1990–2010. *Proceedings of the National Academy of Sciences*, *109*(25), 9775–9780.

Engelbrecht, H.-J. (2009). Natural capital, subjective well-being, and the new welfare economics of sustainability: Some evidence from cross-country regressions. *Ecological Economics*, *69*(2), 380–388.

Falkenstein, E. (2013, July 7). *Stevenson and Wolfers' Flawed Happiness Research*. Retrieved November 18, 2014, from http://falkenblog.blogspot.no/2013/07/stevenson-and-wolfers-flawed-happiness.html

Fleurbaey, M. (2009). Beyond GDP: The quest for a measure of social welfare. *Journal of Economic Literature*, *47*(4), 1029–1075.

Helliwell, J.F. (2006). Well-Being, Social Capital and Public Policy: What's New? *The Economic Journal*, *116*(510), C34–C45. doi:10.1111/j.1468-0297.2006.01074.x

Helliwell, J.F. and Huang, H. (2008). How's Your Government? International Evidence Linking Good Government and Well-Being. *British Journal of Political Science*, *38*(04). doi:10.1017/S0007123408000306

Helliwell, J.F. and Putnam, R.D. (2005). The Social Context of Well-Being. In F.A. Huppert, N. Baylis and B. Keverne (Eds.), *The science of well-being*. Oxford, New York: Oxford University Press.

Huppert, F.A. (2009). Psychological Well-being: Evidence Regarding its Causes and Consequences. *Applied Psychology: Health and Well-Being*, *1*(2), 137–164. doi:10.1111/j.1758-0854.2009.01008.x

Huppert, F.A., Baylis, N. and Keverne, B. (Eds.). (2005). *The science of well-being*. Oxford, New York: Oxford University Press.

Kahan, D.M., Jenkins-Smith, H. and Braman, D. (2011). Cultural cognition of scientific consensus. *Journal of Risk Research*, *14*(2), 147–174. doi:10.1080/13669877.2010.511246

Kahneman, D., Diener, E., Schwarz, N. and Russell Sage Foundation. (2003). *Well-being: the foundations of hedonic psychology*. New York: Russell Sage Foundation.

Kahneman, D. and Krueger, A.B. (2006). Developments in the Measurement of Subjective Well-Being. *Journal of Economic Perspectives*, *20*(1), 3–24. doi:10.1257/089533006776526030

Kahneman, D. and Sugden, R. (2005). Experienced Utility as a Standard of Policy Evaluation. *Environmental and Resource Economics*, *32*(1), 161–181. doi:10.1007/s10640-005-6032-4

Krueger, A.B. and Stone, A.A. (2014). Progress in measuring subjective well-being. *Science*, *346*(6205), 42–43. doi:10.1126/science.1256392

Layard, R. (2011). *Happiness: lessons from a new science,* 2nd ed. New York: Penguin Press.

Leung, A., Kier, C., Fung, T., Fung, L. and Sproule, R. (2013). Searching for Happiness: The Importance of Social Capital. In A. Delle Fave (Ed.), *The Exploration of Happiness* (pp. 247–267). Dordrecht: Springer Netherlands. Retrieved from http://link.springer.com/10.1007/978-94-007-5702-8_13

Lewandowsky, S., Oberauer, K. and Gignac, G.E. (2013). NASA Faked the Moon Landing-Therefore, (Climate) Science Is a Hoax: An Anatomy of the Motivated Rejection of Science. *Psychological Science*, *24*(5), 622–633. doi:10.1177/0956797612457686

Lipton, E. (2014, October 30). *Richard Berman Energy Industry Talk Secretly Taped*. Retrieved November 11, 2014, from http://www.nytimes.com/2014/10/31/us/politics/pr-executives-western-energy-alliance-speech-taped.html?ref=us&_r=2

Lyubomirsky, S., Sheldon, K.M. and Schkade, D. (2005). Pursuing happiness: The architecture of sustainable change. *Review of General Psychology*, *9*(2), 111–131. doi:10.1037/1089-2680.9.2.111

MacKerron, G. (2011). Happiness and environmental quality. PhD thesis. London: School of Economics.

MacKerron, G. and Mourato, S. (2013). Happiness is greater in natural environments. *Global Environmental Change*, *23*(5), 992–1000.

Marshall, A. (1920). *Principles of economics*. Amherst, N.Y.: Prometheus Books.

Mooney, C. (2014, July 22). The Strange Relationship Between Global Warming Denial and …Speaking English | *Mother Jones*. Retrieved October 22, 2014, from http://www.motherjones.com/environment/2014/07/climate-denial-us-uk-australia-canada-english

Pew Research Center. (2012). *Public Priorities: Deficit Rising, Terrorism Slipping*. Pew Research Center. Retrieved from www.pewglobal.org

Putnam, R.D. (2007). E Pluribus Unum: Diversity and Community in the Twenty-first Century The 2006 Johan Skytte Prize Lecture. *Scandinavian Political Studies*, *30*(2), 137–174. doi:10.1111/j.1467-9477.2007.00176.x

Rodríguez-Pose, A. and von Berlepsch, V. (2014). Social Capital and Individual Happiness in Europe. *Journal of Happiness Studies*, *15*(2), 357–386. doi:10.1007/s10902-013-9426-y

Snowdon, C. (2012, June 19). *Shiny happy people? The madness of the Happy Planet Index.* Retrieved November 15, 2014, from http://www.adamsmith.org/research/think-pieces/shiny-happy-people-the-madness-of-the-happy-planet-index/

Stevenson, B. and Wolfers, J. (2008). *Economic growth and subjective well-being: Reassessing the Easterlin paradox.* National Bureau of Economic Research.

Stewart, H. (2012, January 8). *Cut the working week to a maximum of 20 hours, urge top economists.* Retrieved November 12, 2014, from http://www.theguardian.com/society/2012/jan/08/cut-working-week-urges-thinktank

Stiglitz, J., Sen, A. and Fitoussi, J.-P. (2009). *Report of the Commission on the Measurement of Economic Performance and Social Progress (CMEPSP)* (p. 149). Retrieved from http://www.stiglitz-sen-fitoussi.fr/en/documents.htm.

Stoknes, P.E. (2014). Rethinking climate communications and the 'psychological climate paradox.' *Energy Research & Social Science*, *1*, 161–170. doi:10.1016/j.erss.2014.03.007

UNStats. (2012). *System of Environmental-Economic Accounting (SEEA) Central Framework.* UN, EC, WorldBank, IMF, FAO, OECD,. Retrieved from http://unstats.un.org/unsd/envaccounting/White_cover.pdf

Wackernagel, M., Onisto, L., Bello, P., Callejas Linares, A., Susana López Falfán, I., Méndez Garcı a, J., Guerrero, A., Guadalupe Suárez Guerrero, M. (1999). National natural capital accounting with the ecological footprint concept. *Ecological Economics*, *29*(3), 375–390. doi:10.1016/S0921-8009(98)90063-5

Whyte, J. (2013). *Quack science and public policy.* London: Institute of Economic Affairs.

World Bank and Commission on Growth and Development. (2008). *The growth report: strategies for sustained growth and inclusive development.* Washington DC: World Bank on behalf of the Commission on Growth and Development.

CHAPTER 6

The Gap between Attitude and Behavior in Environmental Protection – the Case of Norway

By Caroline Dale Ditlev-Simonsen

INTRODUCTION

Are members of the public "science based activists"? To what extent does improving science based knowledge actually translate into changed behavior in market democracies? When behavior fails to follow knowledge, dissonance increases. This study looks into the issue of blame: Who does the public blame for the perceived and real gap?

The world's combined consumption of resources has only continued to grow since the release of the Brundtland Report, "Our Common Future," over a quarter-century ago. Today, 1.5 Earths would be required to meet the demands humanity makes on nature each year (WWF 2014, Wackernagel, this volume).

People commonly say they are concerned about the environment, prefer environmentally friendly products and are even willing to pay more for them. However, most of us choose the least expensive option when we are shopping (Devinney et al. 2006). The gap between how people believe we should behave and how we actually behave is a central factor in understanding why we are on the wrong path. What is behind this gap? And what can we do to change the pattern? These questions need further exploration.[1]

This chapter is structured as follows: The next section looks at what we know about the gap between attitude and behavior and connects this under-standing to the individual's moral and theoretical perspectives, with a focus on neutralization theory. Section 3 describes five areas – waste disposal, energy conservation, use of public transportation, use of water-saving showerheads, and purchase of organic foods – and presents cases that delineate and docu-ment the gap between attitude and behavior. In section 4 we present a survey conducted in three larger companies, asking who the respondents believe are to blame for this gap. Section 5 discusses the findings and establishes a foun-dation for proposals to reduce the gap.

The study was conducted in Norway, which is an interesting case because the standard of living in the country is one of the highest in the world – a standard many people elsewhere in the world would like to attain. Determining the attitudes and behavior of citizens with this level of welfare might be a good indicator of where other societies are heading. Norway's high standard of living is also reflected in consumption: It has one of the 20 largest ecological footprints in the world. If everyone on Earth lived like an average Norwegian, we would need more than two Earth-sized planets to support them (WWF 2010).

---

1   We would like to extend our gratitude to the UNI Foundation for supporting this study through donations.

6.1 PERSPECTIVES ON ISSUES (DE)LINKING ATTITUDE AND BEHAVIOR?

The industrial world's unsustainable development has led to increased interest in companies' social responsibility, or Corporate Social Responsibility (CSR). *Companies are perceived by many as the major threat to sustainable development because of pollution and their resource use* (Carroll 1999). Companies have responded by taking on social responsibility activity and reporting (Ditlev-Simonsen 2010). Customer and governmental demands for social responsibility on the part of companies have increased significantly, but individuals have not focused on their own environmental and social responsibility with the same intensity. On the contrary, in many ways it seems as if the social and ecological morality of the individual has deteriorated.

THE MORALITY OF THE INDIVIDUAL

The morality of the individual is closely connected to his or her consumer behavior. A range of studies have documented that social and ecological morality at the level of the individual is on the decline.

For example, according to opinion polls conducted by insurance companies seeking to measure the development of insurance morality among Norwegians over the past five years, only 70 percent of respondents in all age groups consider adding a little on top of a claim to be "swindling" the insurance company, as opposed to 81 percent in 2005 (NTB 2007). As many as 16 percent of Norwegian employees surveyed in another study believed it is acceptable to use bribes in the form of cash to expand or maintain an activity (Gedde-Dahl 2009).

In studies conducted in other countries, we see a gap between attitude and behavior. Though 75 percent of Americans regard themselves as environmentalists, such social norms do not easily translate into behavioral changes that have an impact on consumers' choices (Osterhus 1997, Chatzidakis, Hibbert, and Smith 2006). Consumers' social and ecological morality stops at the wallet; they say they care about the environment, but when standing in the shopping aisle they will most likely opt for the cheapest product (Devinney et al. 2006).

Companies that strategize based on market research that shows an increased demand for environmental products have in many cases suffered losses when people don't follow up on what they say. In one international study, 50 percent of those asked replied that they were willing to pay more for organic, environmentally friendly or fair trade products, while another study concluded that the market for ethical products was close to 30 percent. In reality, the market was only a fraction of what the surveys indicated (Devinney et al. 2006).

Other studies confirm that people are not willing to pay an "environmental premium" for green or other environmentally friendly products unless there is some economic benefit from doing so (Michaud and Llerena 2011).

The major problem in achieving sustainable development is that consumption in the developed world is too great. Ironically, the people who exhibit such non-sustainable consumption are those who best understand that it is not sustainable. Every year a significant number of surveys are conducted to examine attitudes about and fears of environmental catastrophes. The results indicate that respondents are concerned about the environment but do not act on these concerns. Thus we see a large gap between what people believe should be done with regard to the environment and what they actually *do*.

COGNITIVE DISSONANCE

How do we explain this gap between attitude and behavior? In behavioral psychology, the concept of "cognitive dissonance" is used to describe situations in which a person does not behave in harmony with what he or she believes is appropriate or perceives as proper behavior. Smoking is a typical example: Nearly everyone today knows that smoking is detrimental to health. Nonetheless, many who smoke continue to do so. Even more surprisingly, many young people *start* to smoke. Most of those who continue to smoke have ready excuses, ranging from "I'm not able to quit" to "The air's so polluted in the city anyway, it makes no difference." Another aspect of cognitive dissonance is the difference between what we actually do and what we *think* we are doing.

Behavioral scientists have found that we *believe* we behave better and more properly than we actually do (Banaji, Bazerman, and Chugh 2003). A similar study – of Norwegians' self-perception – revealed that three out of four men believe they drive better than the average driver (NTB 2009).

Having documented the gap between words and action, how do we explain this cognitive dissonance? The neutralization theory offers a good framework (Chatzidakis, Hibbert, and Smith 2006).

THE NEUTRALIZATION THEORY

Sykes and Matza are recognized as the developers of the neutralization concept. Their theory grew out of an attempt to explain juvenile delinquency (Sykes and Matza 1957). At the time – over 50 years ago – researchers already agreed that what pushed youth into criminality were not physical issues or illnesses, but rather processes and social interaction. Sykes and Matza suggested the following five major neutralization techniques:

▶ *The Denial of Responsibility*
▶ *The Denial of Injury*
▶ *The Denial of the Victim*
▶ *The Condemnation of the Condemners*
▶ *The Appeal to Higher Loyalties*

This justification has been used in many different contexts, but surprisingly little in the study of consumer behavior (Chatzidakis, Hibbert, and Smith 2007). In 2006, Chatzidakis et al. proposed that neutralization theory could also explain the gap between ethical conviction and unethical behavior (Chatzidakis, Hibbert, and Smith 2006).

When applied to ethics and environmentalism, neutralization justifications tend to blend into each other and are difficult to separate. They might be better used to interpret and categorize an individual's statements. The goal of this study is not just to look at *what* justifications people use to explain the gap between attitude and behavior, but more specifically, *who* people blame in seeking to explain them. This approach should provide a better point of departure for establishing initiatives to reduce the gap.

WHICH ACTORS ARE HELD RESPONSIBLE FOR THE GAP BETWEEN ATTITUDE AND BEHAVIOR?

This study focuses on the degree to which blame for the gap is placed on each of the three central actors in society, illustrated in Fig. 6.1: the individual (themselves), the society (people around them) and the authorities. We have chosen not to include companies because in most cases, companies supply what the market demands (such as low-flow showerheads) – what individuals collectively want – and therefore base their activities on the demands of the three central actors. Furthermore, the authorities are the supplier in several of the five areas explored in this study (waste sorting and public transportation, for example). We will look more closely at companies' relationship to environmentally friendly behavior in the last chapter, where findings are discussed.

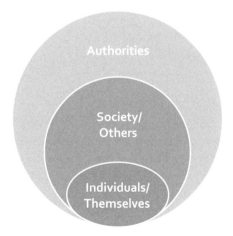

FIGURE 6.1:   Who has responsibility for the gap between attitude and behavior?

BLAME THEMSELVES

The idea that an individual has a sense of right and wrong – a kind of golden mean (virtue) – was proposed over 2 000 years ago by Aristotle. The philosophy is based on the idea that, through continuous training, people can develop a capacity to do *what is right*. A somewhat less "noble" path to proper behavior is related to ethical duty, which is again related to legitimacy. We must (or must not) do certain things that are not directly connected to laws and regulations if we want our behavior to be viewed as acceptable or legitimate. A typical and practical Norwegian example: not eating with your fingers when the table is set with a knife and fork, or not picking your nose in public. There is nothing unlawful about these types of behavior; they are "just not done." We voluntarily place constraints on our behavior that are above and beyond the law.

At the same time, it is acceptable to break certain statutory rules, such as crossing the street against a red light or exceeding the speed limit. This balance brings into play the concepts of duty and virtue, which can help explain individuals' behavior. With respect to the five cases in this study, then, what kind of moral responsibility does the individual have? To what extent does one have a duty to behave in an environmentally friendly way?

The golden rule – do unto others as you would have them do unto you – is a principle of reciprocity found in most religions. The five cases presented in this study all have to do with such a principle. People believe that one *should* participate in waste sorting, energy economization and public transportation and that one should use water sparingly and buy environmentally friendly products. Nevertheless, individuals will find ways to justify not doing so – and this is also acceptable.

The question, then, is to what extent the individual should be held responsible for not behaving in an environmentally friendly way. To what extent does the individual have a moral obligation to behave in such a manner? To what extent is environmental concern something that should come from within as opposed to from without – that is, from society?

BLAME SOCIETY (OTHERS)

People are, to a great extent, group-oriented. As herd animals (Naturen 2008) we follow those around us, which becomes a societal norm. Even when we try to separate ourselves from our fellow humans – to be unique – we become more alike. We follow trends and fashions in behavior as well as in clothes and other consumer products. For example, using a cycling or downhill helmet is not required by law (as opposed to using seatbelts, which is), but more people are using them. This is a trend. Research shows that we allow ourselves to be influenced by society even without knowing it (Goldstein, Cialdini, and Griskevicius 2008).

In the past 30 years it has become increasingly fashionable to behave in an environmentally friendly way. The strong growth in electric cars is an example. In years past, driving around in a big, gas-guzzling American car was something of a status symbol. Today many Norwegians would consider it embarrassing to do so.

We know we are influenced by our surroundings. But how important is it for our surroundings to set a good example – with respect to addressing environmental concerns – and when will this change our behavior? To what extent should society be held accountable for individuals who don't behave in an environmentally friendly way? To what extent do the people around us set the standard for what is proper behavior?

BLAME THE AUTHORITIES

Government authorities are a third actor, beyond the individual and society. The government is responsible for safeguarding collective benefits related to the environment. For example, governments have changed consumption patterns through stricter regulations and pricing mechanisms. Previously, people could dump refuse in the sea and put whatever they wanted in waste containers, but regulations governing special waste and pricing of waste removal have changed these behaviors. Fees and taxes on cars and fuel have led to more energy-efficient cars. At the same time, if initiatives such as these result in higher prices for resources and extra work for consumers, people may react negatively to them.

How much responsibility does the government have when individuals do not behave in an environmentally friendly manner? To what extent do the authorities set the standard for proper behavior?

6.2   THE DEVELOPMENT OF CASES – FIVE CASES THAT DESCRIBE THE GAP BETWEEN ATTITUDE AND BEHAVIOR

In this chapter we present five cases that were developed to describe and illustrate the gap between attitude and behavior. The cases are based on daily activities that everyone can relate to and involve common resources. They are fairly typical for environmentally friendly behavior. Also, they are not associated with any immediate positive consequence for the individual. Instead, the opposite is true: Waste sorting involves more work. In the short term, it is more expensive to install a central heating system. Using public transportation is usually more expensive, time-consuming and less comfortable than using your own car. Low-flow showerheads provide less water pressure and less warmth in the shower. Ecological food is more expensive and the selection is limited. Individuals who change their behavior in these areas to be more environmentally friendly contribute positively to our environment, but it will "cost" them in the short run in the form of inconvenience, constraints or increased expense.

The size of the gap between attitude and behavior, and the environmental effects, vary among the different cases. While using collective transportation instead of a private car has a major impact on the environment, low-flow showerheads have a limited impact. The diversity and widely varying environmental effects of these cases makes it even more interesting to compare them with a common format using the same procedures.

CASE 1: WASTE SORTING (WASTESORT)

In 2010, Norwegians threw away an average of 424 kg of waste (SSB 2011). The amount of waste – both the total amount and the amount per person – has grown continuously since measurements were first taken. Norwegians threw away 70 percent more waste in 2010 than they did in 1995, for example. Twenty years ago, most waste was sent to waste disposal sites. Since 1995, material recovery and biological processing has grown: from approximately 35 percent to 50 percent (Miljøstatus Norge 2011). Source sorting is essential for recycling material, but requires both time and energy on the part of households. According to a survey conducted by the Agency for Waste Management, 63 percent of respondents in Oslo, the capital of Norway, felt that source sorting is one of the most important and effective ways they could contribute to improving the environment. Nonetheless, only 40 percent of households where source sorting had been implemented actually participated in the expanded program. The first case addresses source sorting and documents the gap between attitude and behavior:

> Sixty-three percent of Oslo residents believe it is very important that individual households source sort their waste. At the same time, only 40 percent participate in the expanded program for waste sorting.

CASE 2: ENERGY ECONOMIZATION (ENERGYSAVE)

Energy consumption in Norwegian households is among the highest in the world. Only Iceland uses more energy per person (Bakken 2012). Possibilities for further development of hydroelectric energy are limited. Over 60 percent of Norwegians asked say that, in light of new information about the human contribution to global warming, they will reduce their energy consumption (Elden 2005). At the same time, according to the Norwegian Water Resources and Energy Directorate, interest in energy-saving initiatives is lukewarm.

Central heating systems help reduce electrical consumption because customers can choose sources of energy that are more environmentally friendly than electricity, such as heat pumps, solar heating and earth heat (Hansen and Bjåland 2000). The second case, then, addresses energy economization:

> *Over 60 percent of those asked say that, in light of new information about the human contribution to global warming, they will reduce their energy consumption. Central heating systems help reduce electricity consumption. Nonetheless, only 25 percent of recently completed houses have central heating systems with floor heating.*

CASE 3: USE OF PUBLIC TRANSPORTATION (PUBTRANSP)

Automobiles currently account for 20 percent of the world's CO_2 emissions, and the use of cars continues to grow. Eighty-five percent of all households in Norway have access to a car (Mikkelsen 2008). More than half of Norwegians say they will reduce their energy consumption, and 67 percent of the population is concerned about doing something personally to protect the environment and natural resources (NSB 2011). Taking a train instead of driving or flying can substantially reduce negative impacts on the environment. Traveling by train uses considerably less energy than traveling by car or plane (www.kollektivkampanjen.no). The third case addresses public transportation:

> *Sixty-seven percent of Norwegians are concerned about doing something personally to protect the environment and natural resources. Nonetheless, only 4 percent choose a train when they travel.*

CASE 4: USE OF LOW-FLOW SHOWERHEADS (SHOWERS)

Today, a daily bath or shower is normal in Norway. A bath uses approximately 200 liters of water (7kWh energy). A five-minute shower uses only half that amount: 100 liters. Low-flow showerheads can reduce this amount even further, by half (Enøksenteret 1999).

Energy consumption can be significantly reduced, therefore, by using low-flow showerheads. Makers of showerheads have worked to develop better low-flow technology that saves water without – according to the manufacturers – sacrificing the pleasure of the shower. Still, sales of low-flow showerheads have not taken off. The fourth case addresses low-flow showerheads:

> *Over 60 percent of those asked say that, in light of new information regarding the human contribution to global warming, they will reduce their energy consumption. Nonetheless, low-flow showerheads account for only 25 percent of sales in the private market (despite increased marketing efforts).*

CASE 5: ECOLOGICAL PRODUCTS (ECOFOOD)

Ecological agriculture is founded on the principles of health, ecology, fairness and caution. Ecological foods are characterized by minimal use of additives, attention to the welfare of livestock, and no use of chemical sprays. Norwegians are not concerned about ecological foods, however, so some producers are considering giving up their ecofoods (NTB 2011).

Authorities prefer higher production of ecological food in Norway, so the Ministry of Agriculture and Food has set a goal that, by 2020, 15 percent of production and consumption will be ecological (Landbruks- og matdepartementet 2009). However, ecological foods are on average 40 percent more expensive than conventional foods. While 3 percent of what farmers produce today is ecological, consumers buy only 1 percent ecological (Svartdal 2007). The fifth case addresses ecological products:

> *Forty-four percent of respondents say they are willing to pay more for environmentally and socially responsible products. The government has set a goal that, by 2020, 15 percent of food sales will consist of ecological products. While 3 percent of what farmers produce is ecological, only 1 percent of the food we buy is ecological (the remaining 2 percent is a mix of standard foods).*

Based on these five cases and following the neutralization theory presented in Chapter 2, we will explore what people believe to be the reason for these gaps between attitude and behavior.

6.3   TESTING THE CASES
– SURVEY OF WHO IS RESPONSIBLE FOR THE GAP

Based on the cases above, this chapter will explore who people blame for the gap between attitude and behavior, in order to explain it.

METHODS AND DATA

To investigate how the average Norwegian explains the gap between attitude and behavior, we contacted three large Norwegian companies with a combined workforce of 14 000 employees. We had access to a randomly selected group of 120 to 220 employees from each company, for a total of 540 people, with a response rate of 54 percent.

Two challenges are associated with a survey of such sensitive areas as attitudes and behavior. First, respondents tend to glorify their behavior in their answers. Also, they tend to give answers that are socially but not always factually correct, called an SDR, or "socially desirable response" (Chung and Monroe 2003, Zerbe and Paulhus 1987). To minimize the possibility of SDRs in this study, we avoided asking respondents to evaluate themselves. Instead, we asked them to assess what they think *most people* believe is responsible for the gap between attitude and behavior in the five areas we were investigating. This procedure, called "proxy," is recommended for reducing response bias (Paulhus 1991).

We did not ask about gender, income, age, education or other demographic variables in the questionnaire. The reason for this is twofold. First, previous studies have shown that these parameters have little impact on ethical behavior, which is integrated into an individual's psyche independent of demographic variables (Devinney et al. 2006). Also, excluding demographic questions helps reinforce anonymity. Thus respondents were less likely to filter their answers in accordance with SDR.

On the questionnaire, respondents were asked to cross off, in order, who they blamed for the gap between attitude and behavior: 1) themselves, 2) society or 3) the authorities. Respondents were asked to rank these alternatives on a scale of 1 to 5 (1 = irrelevant, 5 = very relevant). An additional sentence was provided to explain the alternatives. The questionnaire was anonymous and distributed electronically.

RESULTS

The results of the survey are presented in Fig. 6.2. The combined findings show that people blame all three of the actors (themselves, society and authorities), but to varying degrees.

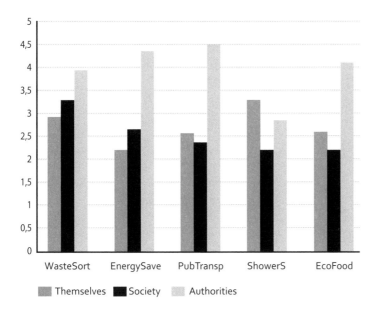

FIGURE 6.2:   Whom the respondents believe people blame in order to explain the gap in the five cases, on a scale of 1 to 5 (1 = irrelevant, 5 = very relevant).

On the whole, respondents blamed the authorities for the gap between attitude and behavior. Thus, people believe that the authorities should step in to reduce the gap. The authorities are regarded as having the greatest responsibility in the area of public transportation (more frequent departures with fewer delays), followed by responsibility for generating initiatives for energy economization (mandatory and less expensive), ecological food (lower prices and better choices) and, finally, waste management (simpler and more practical).

Using low-flow showerheads was the only area where respondents felt people have the most responsibility; i.e. where they blamed themselves. This can be explained by the fact that the authorities are limited in what they can do to make people switch to low-flow showerheads.

With respect to waste management and energy economization, results indicate that society places second, after the authorities, in terms of blame assigned. One possible interpretation is that if it were more common to sort waste and engage in energy economization, more people would do it. An example seems to bear this out: The sorting ratio is greater in single-unit dwellings that have their own waste containers than it is in housing cooperatives with a shared container.

In the areas of public transportation, low-flow showerheads and ecological foods, placing the blame on society – i.e. trends in the population – comes in third. Here, it is assumed we are less influenced by herd mentality.

For waste sorting and energy economization, self-blame comes in third, which means that in these areas respondents believe people blame themselves the least. With respect to public transportation and purchase of ecological foods, self-blame comes in second (public transportation significance $p < 0.20$), which suggests that people take more responsibility for behaving properly.

CONCLUSION

The cases presented here show that people recognize environmental challenges and believe we should adapt our behavior to achieve sustainable development. At the same time, the study shows that people do not always behave as they feel they should. In five cases – waste disposal, energy conservation, use of public transportation, use of water-saving showerheads, and purchase of organic foods – we found that different actors are assigned varying degrees of responsibility for this gap, though the primary blame is placed on authorities. The study found that people do not believe encouraging individuals to voluntarily behave in an environmentally friendly manner will lead to a major shift in consumer behavior. People know what environmentally friendly behavior is about, but they are not inclined to voluntarily take on the negative consequences, such as more work, less freedom, unpleasantness and increased costs.

The study found that people are open to intervention by the authorities to facilitate sustainable development. This finding might reflect individuals trying to excuse their own inactions, or provide an impetus for the authorities to engage in sustainability regulations.

An interesting avenue for further study would be to conduct the same study in countries with different standards of living and political settings.

BIBLIOGRAPHY

Bakken, J.D. (2012). Nordmenn bruker mest strøm i verden. *Nationen*, 04.04.2012, http://www.nationen.no/tunmedia/nordmenn-bruker-mest-strom-i-verden/ Accessed November 21, 2014.

Banaji, M.R., Bazerman, M.H. and Chugh, D. (2003). How (Un)Ethical Are You? *Harvard Business Review, 81*(12), 56–64.

Carroll, A.B. (1999). Corporate social responsibility. *Business & Society 38*(3), 268.

Chatzidakis, A., Hibbert, S. and Smith, A. (2006). 'Ethically Concerned, yet Unethically Behaved': Towards an Updated Understanding of Consumer's (Un)ethical Decision Making. *Advances in Consumer Research, 33*(1), 693–698.

Chatzidakis, A., Hibbert, S. and Smith, A.P. (2007). Why people don't take their concerns about fair trade to the supermarket: The role of neutralisation. *Journal of Business Ethics, 74*, 90–100.

Chung, J. and Monroe, G.S. (2003). Exploring Social Desirability Bias. *Journal of Business Ethics, 44*(4), 291–302.

Devinney, T., Auger, P., Eckhardt, G. and Birtchnell, T. (2006). The Other CSR. *Stanford Social Innovation review* Fall 2006: www.ssireview.org.

Ditlev-Simonsen, C.D. (2010). Historical Account of Key Words in Non-Financial Report Titles. *Issues in Social & Environmental Accounting, 4*(2), 136–148.

Elden, G. (2011). *Egoistisk energisparing?* http://www.forskning.no/artikler/2005/april/1112364544.14. http://www.forskning.no/artikler/2005/april/1112364544.14]. 2005 [cited July 7th 2011]

Enøksenteret (2012). *Enøk i hjemmet – bad eller dusj,* http://www.husogheim.no/1/1_58.html 1999 [cited February 27 2012].

Gedde-Dahl, S. (2009). Ok med bestikkelser, 23.05.2009, side 16. *Aftenposten.*

Goldstein, N.J., Cialdini, R.B. and Griskevicius, V. (2008). A Room
    with a Viewpoint: Using Social Norms to Motivate Environmental
    Conservation in Hotels. *Journal of Consumer Research, 35*(3), 472–482.

Hansen, A. and Bjåland, N. (2000). Spar penger: Vann i gulvet, 18.10.2000.
    VG.

Landbruks- og matdepartementet. (2009). *Handlingsplan for å nå målet om
    15 pst. økologisk produksjon og forbruk i 2020. Økonomisk, agronomisk –
    økologisk!*

Michaud, C. and Llerena, D. (2011). Green consumer behaviour: an
    experimental analysis of willingness to pay for remanufactured products.
    Business Strategy & the Environment, 20(6), 408–420.

Mikkelsen, G.S. (27.02.2012). *Bedre veier NÅ!* http://www.bedreveier.org/
    persontrafikk.html [200827.02.2012].

Miljøstatus Norge (27.02.2012). *Avfall og gjenvinning,* http://www.
    miljostatus.no/tema/Avfall/Avfall-og-gjenvinning/ [201127.02.2012].

Naturen (2008). Vi er flokkdyr. *Illustrert vitenskap,* http://illvit.no/naturen/
    levesteder/vi-er-flokkdyr no. 15:40–47.

NSB (2011). *Personlig kommunikasjon,* October 31, 2011.

NTB (2007). *Flere godtar forsikringssvindel,* http://www.dn.no/forsiden/
    naringsliv/article1197267.ece. (08.10.2007).

NTB (2009). Bilister tror for godt om seg selv http://www.abcnyheter.
    no/nyheter/090714/bilister-tror-godt-om-seg-selv. *ABC nyheter,*
    14.07.2009.

NTB (2011). Nortura vurderer å gi opp økomat, 03.08.2011. *Aften,* 10.

Osterhus, T.L. (1997). Pro-social consumer influence strategies: When and
    how do they work? *Journal of Marketing, 61*(4), 16.

Paulhus, D.L. (1991). Measurement and Control of Response Bias. In
    Measures of personality and social psychological attitudes, edited by J.P.
    Robinson, P.R. Shaver and L.S. Wrightsman, 17–59. San Diego, CA:
    Academic Press, Inc.

SSB (2012). *Mindre avfall til deponi,* http://www.ssb.no/emner/01/05/10/
    avfkomm/ 2011 [cited February 27 2012].

Svartdal, T. (2012). *Økologisk mat – for dyrt?,* www.naturvernforbundet.no
    2007 [cited 09.09 2012].

Sykes, G.M. and Matza, D. (1957). Technizues of Neutralization: A theory
    of Delinquency. *American Sociological Review, 22*(6 (Dec.)), 664–670.

WWF (2012). *Living Planet Report.* http://www.wwf.no/dette_jobber_
    med/norsk_natur/naturmangfold/okologisk_fotavtrykk/ 2010 [cited
    February 29 2012].

WWF (2014). *Living Planet Report 2014.* edited by http://wwf.panda.org/ about_our_earth/all_publications/living_planet_report/.

Zerbe, W.J. and Paulhus, D.L. (1987). Socially Desirable Responding in Organizational Behaviour: A Reconception. *Academy of Management Review, 12*(2), 250–264.

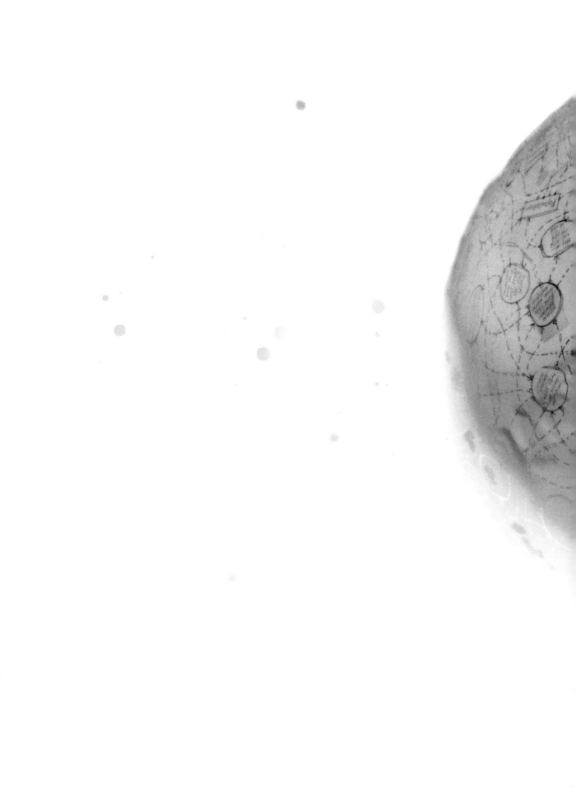

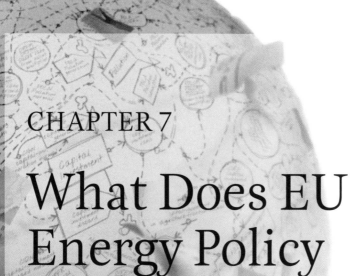

CHAPTER 7

What Does EU Energy Policy Mean for the Climate?

By Nick Sitter

INTRODUCTION

Although energy policy was written into the Treaty of Rome, and has therefore been on the European Union's (EU) agenda since it was created as the European Economic Community (EEC) in 1958, there was no real energy policy to speak of until well into the 1990s. For understandable reasons, the EEC member states decided to leave energy policy as a national matter. The Treaty of Rome was negotiated shortly after the 1956 Suez crisis, and the next two decades saw another two wars in the Middle East, the emergence of the Organization of the Petroleum Exporting Counties (OPEC), and the oil price hike after OPEC's embargo of several Western states. All this took place at the same time as the EEC member states kept their gas and electricity markets firmly in the hands of state-owned national monopolies.

It was only with the end of the so-called "OPEC era" in oil, the establishment of the Single European Market (SEM), and the gradual liberalization of gas and electricity markets in the 1990s, that EU energy policy came of age. Interestingly for the question that this chapter addresses, this happened at the same time as environmental policy was added to the Treaty. Since then, EU energy policy has had three central objectives: to establish a liberalized single European market in energy, to ensure the security of energy in the sense of stable supply at an affordable cost, and to ensure that all this is done in a way that is compatible with environmental protection. The central question in this chapter is to what extent the three objectives have turned out to be compatible: does meeting one goal make the other more obtainable; do these goals involve tradeoffs, or are they incompatible, and what this has meant for the climate?

7.1   THE TRIPLE-BOTTOM LINE OF EU ENERGY POLICY

European Union energy policy – in the sense of energy policy that is made or coordinated at the EU-level, rather than simply the sum of the member states' energy policies – was born at the same time as the SEM. It is therefore no coincidence that EU energy policy is built on market principles, with a central focus on the establishment of competition-based markets in Europe. This "liberal paradigm" has informed EU policy both at home and abroad, and has shaped both the principles behind the policy and the policy tools with which it is pursued (Andersen and Sitter 2009; Goldthau and Sitter 2014). The central idea is that competitive markets complement environmental policies. Effective and transparent markets go hand in hand with policies based on estimating the cost of pollution and meeting environmental policy goals in a cost-effective manner, and at a cost that is politically and socially acceptable (Talus 2013).

The fundamental principle behind EU energy policy is a liberal, rule-based, approach to markets. The EU, or more precisely the EEC, was established by six West European states as one of several international organizations designed to promote prosperity and democracy. The most important mechanism whereby this was to be achieved was economic integration, and the tools with which the new institutions were equipped were primarily those of regulation (Majone 1996). The European Commission was charged with implementing policy or overseeing policy implementation, and its strongest tool was competition law (McGowan and Wilks 1995). In the 1960s and 1970s the organization's main task was breaking down barriers to trade. With the establishment of the SEM this expanded to include "flanking policies" such as social policy and environment protection, as well as the extension of the SEM to new sectors such as the gas market. The common theme that runs through these efforts is that the Commission interprets most of the problems it faces in terms of market failure. This shapes its approach to tackling new challenges, whether in the form of integrating energy markets, securing the supply of energy, or ensuring that the energy policies of the member states and the EU as a whole do not unduly damage the environment.

The policy tools of the EU are very much the policy tools of a regulatory state. This holds both for the organization as a whole and for its executive arm – the European Commission – in particular. The EU is a regulatory state by design, both in terms of its mandate and its policy tools (Majone 1996, Lodge 2008). Its core mandate is to build markets, and to make them work better. Its policy tools include the power to pass legislation in most areas linked to the SEM, to coordinate member state policy, and to gather and disseminate information. But the EU does not have the two other most important sets of policy tools available to states (Hood 1983): it controls only very limited economic resources, and cannot exercise power directly through public organizations or nationally owned firms. The reach of the EU regulatory state was somewhat limited in the 1970s and 1980s, particularly in terms of energy, since the member states operated their utilities as national (or regional) monopolies. However, with the wave of privatization and liberalization that the member states embarked on in the late 1980s and early 1990s, the time of the regulatory state had come. Indeed, with the end of the Cold War and what Fukuyama memorably labeled the "End of History" (1989), the EU regulatory state grew strong in a context that was remarkably benign, both at home and internationally.

All the three core goals of EU energy policy were therefore elaborated and operationalized during the heyday of the regulatory state. They were very much shaped by liberal principles and regulatory tools. Market integration, security of supply and sustainable consumption were framed in a common paradigm: as a matter of making markets and correcting market failures. This involved establishing common goals and priorities for EU-level policy, and making sure that the EU-regime was compatible with the considerable national diversity in terms of energy production, import and consumption. New EU-level legal regimes and policy tools were developed, and efforts were made to ensure that the policies aimed at building a competitive market were compatible with security of supply and protecting the environment. Successive white and green papers (Commission of the European Communities 1988, European Commission 2000, 2008) elaborated these three goals.

The first objective of the EU's comprehensive energy policy was extending the SEM to the energy sector. In practice, this meant liberalizing electricity and gas markets. The oil market had functioned as an international commodities market for decades, and since the mid-1980s "counter-shock" oil prices had been relatively low because of a combination of high levels of global supply and more efficient consumption. Power production remained a matter for the member states, each of which chose very different strategies in terms of the mixture of fossil, nuclear and renewable sources. But trade in gas and electricity was largely based on contracts linked to long-term projects, rather than competitive commodities trade on spot markets.

The Commission's first effort to change this was greeted with horror from the member states. They sent its proposal for a directive liberalizing electricity and gas markets straight back to the sender (Andersen 1993). Consequently, after receiving instructions from the member states, the Commission developed two separate proposals for the liberalization of electricity and gas markets. The directives were adopted in 1996 (electricity) and 1998 (gas), with follow-up directives for the gas sector in 1998 and 2003. In both cases the Commission saw its liberal principles triumph, but at the cost of what might be called "fuzzy integration" (Andersen and Sitter 2009).

Both industries were prone to market failures: they featured natural monopolies in the form of transmission networks (which required regulation of access, ownership and tariffs), and large (former monopoly) incumbent operations. In the gas sector, state-owned monopoly exporters from Russia and Algeria shaped the broader regional market. The result was a compromise that was liberal in principle, but permitted the member states considerable room for discretion in terms of how they implemented the SEM rules. The member states' authority to control the energy mix and use of natural resources was eventually written into the EU treaty of Lisbon in 2009. The treaty introduced a new article dealing directly with energy, which states that "[these] measures shall not affect a Member State's right to determine the conditions for exploiting its energy resources, its choice between different energy sources and the general structure of its energy supply" (article 194).

The second overall objective was security of supply. Although security of energy supply can be defined simply as stable supply at acceptable prices (Yergin 2006), this has very different implications for oil, gas and electricity markets. For oil, an internationally traded fungible product, the primary risk comes in the form of price volatility. The EU does not have the tools to do much about this on its own, but has joined forces with other consumer states in the International Energy Agency. It is committed to keeping reserve oil stock, and coordinating the release of these in the event of a crisis, as well as to improving information about global supply and demand. For electricity markets, security of supply in the 1990s was regulated very much in terms of public service (including affordable prices and obligations to supply customers). This was taken care of by permitting the member states to maintain national regulation on the matter. It was even written into the EU Treaty with the 1997 Treaty of Amsterdam in the form of Article 7.

However, it is in the gas sector that security of supply has proven most controversial. This is because of the role played by external suppliers. Successive Ukrainian crises in 2006, 2009 and 2014 illustrated the danger of a gas supply cut-off in the event of political conflicts outside the EU. Internally, the different pricing and contracting strategies Gazprom used towards different member states caused concern that the Russian monopoly exporter might be abusing its dominant position, in violation of EU competition law. And the "pipeline diplomacy" between the EU and Russia over the EU's effort to establish a "southern corridor" to access gas from the Caspian Sea demonstrated that regional gas trade has a strong geopolitical dimension.

The third objective of EU energy policy is its environment dimension. Environment policy was first developed at the EU-level in the context of the Single European Act. It was driven by German and Danish concerns that the combination of southern enlargement to Greece, Portugal and Spain in the 1980s might undermine their national environment policies. Since then, it has become an integrated part of several other policy sectors, and emerged as a high-profile issue for the (directly elected) European Parliament. As in the case of completing the SEM and securing supply, the Commission approached the challenge primarily as a matter of correcting market failures.

Whether the issue is disposal of oil platforms (Löfstedt and Renn 1997) or regulation of car engines and fuels (Friedrich, Tappe and Wurzel 2000), if the Commission proposes new rules it does so with a view to correcting, mitigating or preventing market failures. Since the SEM was established, climate change has emerged as the world's single most important negative externality: the biggest market failure of all (Stern 2005). The main problem for the European Commission is whether the EU's environmental policy priorities and the policy tools it uses to pursue these can be made compatible with its pursuit of integrated electricity and gas markets and security of energy supply.

Are the three elements in what might be labeled the EU's "triple bottom line" in energy policy actually compatible? The Commission's overall view is that they are. All three policy objectives and their tools are informed by a single public policy paradigm: identify market failures and correct them. However, several studies have documented disputes and conflicts over goals, priorities and policy tools, particularly in terms of liberalization vs. the interests of the energy industry (Ferman 2009) and industry interests and Commission initiatives on emissions trading and renewables (Dreger 2014).

EU energy policy affects the climate in two ways: directly, in the sense that EU energy policy decisions affect the factors that cause climate change; and indirectly, because EU energy policy affects the EU's (and its member states') policies designed to combat climate change. The next two sections address the direct and indirect effects of EU energy policy in turn. In each case, the question is whether the EU's energy policy goals are compatible with the efforts to combat climate change (or if the two policies even reinforce each other); whether the two policy goals involve trade-offs (achieving more of one comes at the cost of achieving less of the other); or whether the EU is confronted with genuine policy dilemmas in the sense that the two goals are incompatible and they requires choices that involve downgrading or giving up one of the two.

7.2   THE DIRECT EFFECT: THE EU REGULATORY STATE AND THE ULTIMATE EXTERNALITY

Pollution – and ultimately climate change – is one of the most famous examples of how voluntary economic transactions between two actors can have a negative impact on third parties. From a political economy perspective, the normative reason for government action to combat climate change is that it is an externality of extraordinary proportions – one that even affects third parties across borders and generations. The scientific and policy debates about the nature and implications of man-made climate change notwithstanding, the issue in the EU today is not whether the consumption of energy causes climate change. The issue is how to mitigate the effect of the EU's policies on energy production, trade and consumption on the climate.

The first and most important question concerns the impact of the EU policy regime on the energy mix. Since decisions about the energy mix are

firmly located at the member state level, the question about the effect of policies in the EU on energy production affect the climate merits analysis of the
28 member states (or 31 if Norway, Iceland and Liechtenstein are included,
as members of the European Economic Area). This is beyond the scope of
this chapter. As for EU-level policy, the two main questions are: does the EU
policy of leaving the energy mix up to the member state exacerbate climate
change or hinder the EU's efforts to combat climate change; and does the
EU's efforts to coordinate the states' policies on renewable energy contribute
significantly to combating climate change?

The answer to the first question is that the heterogeneity of national energy
markets does indeed permit some member states to maintain commitments to
the use of energy sources that are particularly damaging in terms of climate
change, such as coal. However, here EU policy is a consequence of member
state diversity, not the cause of it. Indeed, the battles over electricity and
gas liberalization in the 1990s demonstrated very clearly that for many states
energy is a "strategic good." It is of vital importance not just to industry, but
to national security. Therefore, any efforts to harmonize policy on the energy
mix would be futile. The EU remains a political system designed to manage
heterogeneity rather than eliminate it (Andersen and Sitter 2014).

The most significant exception to the rule that the member states control
the energy mix is the EU's commitment to increasing its share of renewable
energy. The 2009 Climate and Energy Package committed the EU, through
a legally binding directive, to the triple "20-20-20" targets for 2020: a 20%
reduction in EU greenhouse gas emissions from 1990 levels; raising the share
of EU energy consumption produced from renewable resources to 20%; and
a 20% improvement in the EU's energy efficiency. This is also the central
pillar in the EU's internal efforts to combat climate change. It is reflected in
the national governments' consensus on this policy goal (at least before the
global economic crisis: the national EU leaders reached political agreement
on the 20-20-20 goals in 2007). The states have agreed that this policy priority should be coordinated at the EU-level to overcome a collective action
level, and allowed the Commission to take the lead (Jordan, van Asselt, Berkhout, Huitema and Rayner 2012).

The second energy policy question that has a direct effect on climate
change is the Commission's effort to establish a fully functioning SEM for
electricity and gas. This effort involves three important elements. All else
being equal, a fully integrated energy market, with gas traded on spot markets
rather than bought by utilities at the border, should bring lower prices and
consequently higher consumption. With energy taxation being a matter for
the states, and therefore not necessarily linked to lower commodities prices,

the effect hinges on how the price gains are shared by transmission operators, distributors and consumers. If the introduction of gas-on-gas competition in Germany is anything to go by, the effect might well be lower costs for both industrial and household consumers (Stern and Rogers 2012).

However, all else is rarely equal, and certainly not in the case of EU energy markets. The second element in the equation is whether cheaper gas prices can and will lead to a bigger role for gas at the expense of coal. Again, national policies and subsidies come into play. Finally, the SEM policy regime has been accompanied by both EU and national measures to improve energy efficiency and increase the share of renewables. The answer therefore depends on whether the SEM in energy can be designed in such a way as to reduce rather than increase pollution. The fact that the EU represents a mere 11.5 percent of total global annual carbon emissions, and that this share is shrinking, reflects both the increasing energy efficiency of production and the shrinking share of the EU in the global economy (BP 2013).

The third aspect of EU energy policy that directly affects climate change is the organization's response to the problem of energy security. As of 2012, more than half (53.4%) of the EU-28's gross inland energy consumption came from imported sources (Eurostat 2013). The figures for oil (84.9% imports) and gas (66.7% imports) are even higher. Because oil is traded internationally, on spot markets, most shocks have an effect on price rather than on availability. Gas, however, is traded regionally and bilaterally. This leaves importing states vulnerable to interruptions in supply in the event of external shocks. The EU's main focus in terms of security of supply has therefore been on making sure gas markets are robust, for example by improving interconnection between national markets (including technical measures to ensure that gas can flow both ways in a pipeline), and diversifying external sources of gas supplies and transit routes.

Since Russia has a common interest with the EU in developing alternative pipeline routes (until the Nord Stream pipeline directly to Germany under the Baltic sea was opened in 2012, 80% of all Russian gas to the EU crossed the Ukraine), pipeline diversification has proven more easily achieved than diversification of sources. Norway's share of the EU gas market now rivals that of Russia, but negotiating access to Azeri and Turkmen gas has proven far more elusive for the EU (Goldthau and Sitter 2014). The key question in terms of climate change is to what extent the EU's effort to reduce its vulnerability to interruptions of gas supplies affects the energy mix. If gas maintains its attraction in terms of both price and reliability it may reduce the demand for other fuels, both those that are more harmful (coal) and those that are better from a climate perspective (renewables).

These three energy policy issues – the energy mix, the internal market and the security of external supplies – are part and parcel of an overall security of supply policy. The short-term security of supply challenge concerns price volatility in oil and interruption of gas supplies, the medium-term concern is that energy prices are efficient and affordable, and the long-term element of secure supply concerns sustainability. The overall goal can be coherent even if the individual policy tools involve trade-offs. Measures designed to improve energy market transparency, cope with oil shocks and diversify imported gas are all compatible with the goal of medium-term efficient pricing and a longer-term commitment to a shift away from a carbon-based energy economy.

In the short- to medium-term the three aspects of the EU's internal energy policy are reasonably compatible with the EU's efforts to combat climate change. The main challenge comes in the shape of the national energy mix, and concerns the extent to which the member states can remain committed to renewables and emissions targets at times of global economic crises. The potential problem lies not in the efforts to build a common gas and electricity market, but rather in the loopholes that these projects entail. However, in the longer term, the answer to what the EU's energy policies mean for the environment hinges on whether gas has a role to play as a bridging fuel. Can it make energy transition more realistic and affordable, or might a bigger role for gas merely serve to extend fossil fuel dominance. To the extent that the relatively high cost of renewables and lower oil prices persist, the "bridging fuel" argument seems likely to triumph.

7.3   THE INDIRECT EFFECT:
POLICY COHERENCE AND CLIMATE CHANGE

The EU has made combating climate change one of its top priorities on the international stage. It has taken a lead role in arguing for climate targets, it has pushed for common global norms and guidelines, and it has promoted innovative policy instruments such as its own Emissions Trading Scheme (ETS). Indeed, this form of "experimental governance" is part of a broader pattern of EU policy-making, whereby the EU sets overall goals and metrics, the member states work out plans for meeting these, and the results are subject to reporting, monitoring periodical policy revision (Sabel and Zeitlin 2008). The EU's approach to climate change at the international level is largely shaped by the liberal paradigm that guides the EU at home: the EU uses regulatory tools at home and advocates rule-based international regimes at the global level (Goldthau and Sitter 2014).

The European Commission and the European Parliament both demand EU action on climate change, as they have long done on environment policy in general. Most of the member states support this demand. On the supply-side, the Commission has therefore been able to play the part of a policy entrepreneur. At times it has gone far beyond what the member states were prepared to accept, for example in the form of its proposals for CO_2 taxation. But successful Commission initiatives include the ETS and the EU's 20-20-20 target for increasing the share of renewable sources of energy in the energy mix. However, the big question that haunts all EU external action – whether its policy initiatives in different sectors are coherent – also applies to the organization's efforts to combat climate change (Niemann and Bretherton 2013, Peters 2015).

The internal dimension of the EU's climate policy centers on efforts to reduce pollution and emissions, and increase the share of renewables in the energy mix. Although some of its policy proposals have met with fierce resistance from industry and member states, they have by and large proven compatible with EU energy policy and the SEM. For example, the Commission's car emission policy proposals (the Auto-Oil I and Auto-Oil II programs) saw the car and petroleum industry fight over whether the main burden should fall on engine improvement or petrol quality, while the European Parliament and the member states fought over the overall limits on emissions (Friedrich, Tappe and Wurzel 2000).

In the 1990s, the Commission pushed for EU rules on CO_2-taxation. The idea was to develop a carbon-tax regime at home that could serve as a model for international action. However, because all legislation on taxation required the unanimous consent of the member states, the project was eventually abandoned. The ETS was the compromise outcome: a trading scheme for emissions permits that was launched in 2005 and has been subject to several subsequent revisions. Again the aim was not merely to act at home, but also to develop a regime that could be used as a model for international action (Vogler 2009).

In terms of its internal action on pollution and climate change, EU energy and environment policy initiatives have involved some very real trade-offs. Both the energy-producing and energy-intensive industry have found the Commission's Directorate Generals (DGs) responsible for industry and energy more receptive to their arguments, whereas the DGs responsible for competition and the environment have been less sympathetic to the industry's claims for exemptions. The results have been the kind of compromises that characterize EU energy policy as a whole. The Commission's liberal paradigm has triumphed, and this includes a commitment to combating climate change as a big market failure. But as the successive revisions of gas market legislation and the ETS illustrate, this has been a gradual process of compromise and trade-offs. The EU's policy compromises involve trade-offs with member state energy policies, and between the policy priorities of the Commission's different DGs.

At the international level, the main problem from the EU's perspective has been that the kind of rule-based international regimes it prefers involve a collective action problem that may prove insurmountable at the global level. The simple problem is that climate change (or rather, limiting it) is a "common pool" resource: all states have access to the benefits, whether they exercise the restraint that is required to secure sustainable development or not. However, as Elinor Ostrom famously concluded (1990: 216), overcoming common pool resource problems does not necessarily require "an omnipotent entity called 'the government,'" but it does depend on the main users of said resources being capable of replacing a focus on maximizing short-term interest with "long-term reflection about joint strategies to improve joint outcomes."

The rise of emerging economies as significant carbon emitters means that any solution must involve countries like China and India in addition to the EU and the USA. The EU has consistently pushed for the expansion of international environmental law in order to promote sustainable development (Kelemen and Vogel 2010). In light of Ostrom's theory, the joint US-China initiative to limit emissions announced in November 2014 (*Financial Times*, 14 November 2014) might indicate a more realistic approach to the problem than the global regime favored by the EU.

There is little conflict between the EU's international energy policy and its global approach to climate change. The EU's policy largely amounts to an effort to build international markets and to make whatever regimes exist at any given time work better. In practice this means a commitment extending the rule of the World Trade Organization to cover energy, and to build a new regime for energy trade, transit and investment. To the extent that such efforts fail, the fallback position is to work with the IEA to improve energy market transparency, and for the EU to apply its own regulatory regime to regional gas trade. These policy tools are broadly neutral with respect to the EU's climate initiatives, such as taking a lead role in the process leading up to the Kyoto Protocol, advocating a post-Kyoto agreement, and pledging increased unilateral emission cuts if others follow suit. For example, the 20-20-20 targets came with a conditional offer to reduce emissions further (up to 30 percent) if developing countries implement comparable policies.

In short, the EU's energy policy entails no significant obstacles for the EU's approach to combating climate change on the international level. However, as the debate about the cost of the German Energiewende (the energy transition toward renewables launched in 2000) shows, the really tough trade-offs are likely to be fought out at the national level rather than in Brussels (*Financial Times*, 25 November 2014).

Ford Prefect, the protagonist of Douglas Adams' *The Hitchhiker's Guide to the Galaxy,* spent seven years studying the Earth only to revise the entry in the eponymous guide from "harmless" to "mostly harmless." Much in the same vein, this inspection of the EU's approach to energy policy and climate change points towards the conclusion that the EU's energy policy is *mostly harmless.* But the modifier "mostly" is important. The most significant potential for serious policy conflicts between energy and environment policy lies at the member state level, and arises when policies designed to combat climate change are at odds with industrial policy initiatives. Before the global financial crisis broke out in 2008, much was made of the scope for green growth. The calls for a low-carbon recovery from the crisis continue to echo this, but on a smaller scale. So EU energy policy involves challenges for climate change policy simply because it institutionalizes such a variety of member state energy policies – and some of these national policies are bound to be more incompatible with the climate change agenda than others.

As far as EU-level policy is concerned, the central idea in the EU is that of a well-functioning, competitive, transparent market with reasonable security of supply, and that this can contribute to the EU's efforts to combat climate change. However, EU energy policy in fact turns out to be more or less neutral in terms of climate change. The claim that the SEM provides a good framework for experimenting with innovative policy instruments for reducing pollution is hardly the same as claiming that EU energy policy actually helps mitigate climate change. The first part of the answer to the question "what does EU energy policy mean for the climate" is therefore that EU energy policy provides a stable, rule-based, long-term policy framework within which concerted EU action to combat climate change is possible.

The second part of the answer is that the EU energy policy regime positively contributes to the EU and its member states' climate change initiatives because the SEM in energy has been designed and developed with a view to sustainable energy policy. The strongest part of this claim rests on the 2009 energy and climate package, and particularly the legal commitment all member states have made to the 20-20-20 goals. This is designed to overcome a collective action problem – precisely the sort of public policy provision that the EU was invented to deal with. Even here, however, much depends on how the individual states choose to meet these targets. To the extent that gas replaces dirtier fossil fuels and can contribute to bridging a gap before renewables become competitive in terms of price and reliability, this aspect of energy policy contributes to the EU's overall climate policy agenda. Finally, and perhaps more controversially, given that the EU's approach to dealing with pollution is market-based, a well-functioning and transparent market may well be a necessary condition for a transition to a low-carbon economy.

The third, and final, part of the answer is that the EU's domestic energy and climate policy initiatives provide an arena for experimenting with new policy instruments. This, in turn, might provide models for international climate policy regimes. Like the EU regime, almost all imaginable realistic international regimes must involve rules, regulations and common targets, rather than other policy instruments such as expenditure or taxation, and they must leave much of the question of how to meet targets to sovereign states. Although the EU's market-based approach to energy and climate policy has attracted much criticism, its potential impact as a model for international operation should not be dismissed out of hand.

In short, what does EU energy policy mean for the climate? EU energy policy is broadly compatible with the climate change agenda, but it certainly does not eliminate conflicting policy goals at the national level. EU energy policy involves a number of initiatives that contribute to combating climate change, but it leaves practical implementation to its member states. EU energy and climate policy has involved experimentation with innovative policy tools, but these have so far not proven attractive at the global level. Perhaps the clearest part of the answer is that EU energy policy might not have done much for the climate (yet), but it offers some indications of how climate change might be addressed, at the national, regional and global level.

BIBLIOGRAPHY

Andersen, S.S. (1993). Energy Policy: Interest Interaction and Supranational Authority. In *Making Policy in Europe: the Europeification of National Policy-Making*, edited by S.S. Andersen and K.A. Eliassen. London: Sage.

Andersen, S.S. and Sitter, N. (2009). The European Union Gas Market: Differentiated Integration and Fuzzy Liberalization. In *Political Economy of Energy in Europe*, edited by G. Fermann. Berlin: BWW.

Commission of the European Communities (1998). The internal energy market. *Commission working document* COM (88) 238 final (2 May 1988).

BP. (2013). *Statistical Review of World Energy*. London.

Dreger. J. (2014). *The European Commission's Energy and Climate Policy: A Climate for Expertise?* London: Palgrave.

European Commission (2000). *Green Paper: Towards a European strategy for the security of energy supply.*

European Commission (2008). *Second Strategic Energy Review – an EU Energy Security and Solidarity Action Plan.*

Eurostat (2013). *Energy, transport and environment indicators.* Luxembourg: Publications Office of the European Union.

Fermann, G. (ed.) (2009). *The Political Economy of Energy in Europe: Forces of Integration and Fragmentation,* Berlin: Berliner Wissenschafts-Verlag.

Financial Times (14 November 2014). China and US deal to curb emissions draws mixed response.

Financial Times (25 November 2014). The growing absurdity of German energy policy.

Friedrich, A., Tappe, M. and Wurzel, R.K.W. (2000). A new approach to EU environmental policy-making? The Auto-Oil I Programme, *Journal of European Public Policy,* 7(4), 593–612.

Fukuyama, F. (1989). The end of history. *The National Interest, 16*(Summer), 3–18.

Goldthau, A. and Sitter, N. (2014). A Liberal Actor in a Realist World? The Commission and the External Dimension of the Single Market for Energy. *Journal of European Public Policy, 21*(7), 1452–72.

Hood, Ch. (1983). *The Tools of Government.* London: Macmillan.

Jordan, A., van Asselt, H., Berkhout, F., Huitema, D. and Rayner, T. (2012). Understanding the Paradoxes of Multilevel Governing: Climate Change Policy in the European Union. *Global Environmental Politics, 12*(2), 43–66.

Kelemen, R.D. and Vogel, D. (2010). Trading Places: The Role of the United States and the European Union in International Environmental Politics. *Comparative Political Studies. 43*(4), 427–456.

Lodge, M. (2008). Regulation, the Regulatory State and European Politics. *West European Politics, 31*(1/2), 280–301.

Löfstedt, R. and Renn, O. (1997). The Brent Spar controversy: an example of risk communication gone wrong. *Risk Analysis, 17*(2), 131–136.

Majone, G. (1996). *Regulating Europe.* London: Routledge.

McGowan, L.E.E. and Wilks, S. (1995). The first supranational policy in the European Union: Competition policy. *European Journal of Political Research, 28*(2), 141–169.

Niemann, A. and Bretherton, C. (2013). EU external policy at the crossroads: The challenge of actorness and effectiveness, *International Relations*, *27*(3), 261–275.

Ostrom, E. (1990). *Governing the Commons: The Evolution of Institutions for Collective Action*. Cambridge: Cambridge University Press.

Peters, I. (ed.) (2015). *The European Union's Foreign Policy in a Comparative Perspective*, Berliner Wissenschafts-Verlag.

Sabel, C.F. and Zeitlin, J. (2008). Learning from Difference: The New Architecture of Experimentalist Governance in the European Union. *European Law Journal, 14*, 278–80.

Stern, J. and Rogers, H. (2012). The Transition to Hub-Based Gas Pricing in Continental Europe. In *The Pricing of Internationally Traded Gas*, edited by J. Stern. Oxford: Oxford University Press.

Stern, N. (2005). *Stern review on the economics of climate change*. London: Cambridge University Press.

Talus, K. (2013). *EU Energy Law and Practice: A Critical Account*. Oxford: Oxford University Press.

Vogler, J. (2009). Climate change and EU foreign policy: The negotiation of burden sharing. *International Politics, 46*, 469–490.

Yergin, D. (2006). Ensuring Energy Security. *Foreign Affairs, 85*(2), 69–82.

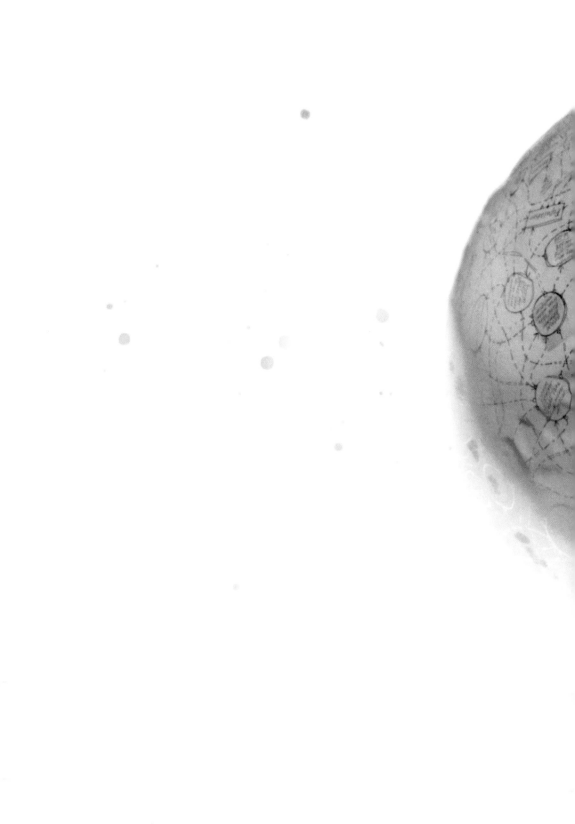

CHAPTER 8

The EU's Climate and Energy Policy – A Case Study of the Adoption of the Climate Change Package in 2008

By Kjell A. Eliassen,
Marit Sjøvaag Marino and Pavlina Peneva

INTRODUCTION

Are democratic institutions suitable for solving the climate challenge, or are they, as often has been claimed, incapable of adopting and implementing effective climate policies simply because such effective policies generally are (a bit) more expensive than the cheapest option in the short run? Are there inevitable trade-offs between policy efficiency, accountability and democratic governance, and if so, how much weakening of democracy and accountability should we tolerate in order to meet the climate challenge? Is the political process in the EU sufficiently removed from the electorate to increase the probability of reaching the ambitious targets from 2008 and subsequent, even more stringent, targets?

The European Union had by the Copenhagen summit in 2009 become one of the leading actors globally in developing and implementing climate change policies. The lack of engagement from the US during the Bush administration gave the EU the possibility to become the leading power in global climate change, to combat and make climate policy an important part of an enhanced external policy of the Union (Oberthur 2009). However, after 2004 the Union included new member states with a very different willingness – and ability – to participate in a combined policy to reduce their emission of greenhouse gases. Despite such large differences between the Member States, the EU was able to create an agreement on a programme of joint action to combat climate change among all of its (at the time – before Croatia) 27 member-states. This achievement is impressive, and even more so when we take into consideration that individual member states had found it difficult to attain political agreement, let alone implementation, on such a programme domestically.

In this chapter we argue that the possibility of coaxing less ambitious member states into agreement through internal burden-sharing arrangements was central to the 2008 climate policy package. We are here in agreement with other analysts. But in addition to this "buying off" of less willing states, we also claim that the political capital invested through the political leadership of the presidency, particularly of German Chancellor Angela Merkel and French President Nicolas Sarkozy, and later Swedish Prime Minister Fredrik Reinfeldt, was a *sine qua non* for the successful outcome of the negotiations.

The conclusion we draw is therefore that democratic systems do have a potential to deliver ambitious targets for climate policies, and that investment of political capital, the benefits of which are reaped not only through immediate re-election but also through credibility and institutionalization of political goals, will contribute to increasing the probability of implementation of such goals.

In our empirical study we focus on three issues:

1. How did the European Commission, as a non-elected executive, prepare for the climate package (e.g., the composition of the package, policies included, legal instruments, calculation of member states' costs)?
2. To what extent and how did the existence and functioning of the Emissions Trading System (ETS) play a role in the success of the Commission's efforts?
3. What was the role of the German EU Presidency in putting this process in motion and later on the French Presidency in concluding this deal, and to what extent and how did the Commission use the role of the Presidencies for these purposes?

This chapter uses the case of the role of the European Commission in the adoption of the 2008 Climate Change Package (often referred to as 20-20-20) to investigate the role of individuals, democratic capacity and political capital in climate policy. The scope of this chapter is restricted to the political process leading up to the adoption of the package (i.e. 2006–2008), rather than giving an up-to-date report on EU climate policy. We focus on the European Commission, and on the interplay between the European Commission and the (primarily) German and French EU Presidencies. In this way we hope to be able to draw some at least tentative conclusions about possibilities and limitations of the political process in the EU.

8.1   THEORETICAL FRAMEWORK – ACCOUNTABILITY AND POLICY EFFICIENCY IN MULTI-LEVEL GOVERNANCE

The classic model of accountability in democratic systems assigns the electorate with a "control function", so that if politicians do not deliver on their electoral promises, they will not be re-elected.

Legitimacy of political decisions can be achieved *ex ante*, by ensuring that the decisions are taken according to some specified procedure, or *ex post*, by the outcome of the decision pleasing at least a majority of the affected constituency.

Although recently elevated to a "separate" policy field with its own DG, climate policy has traditionally been seen as part of environmental policy. After its inclusion in the single European Act, environmental policy has become one of the most rapid developing sectors of EU regulation (Princen 2009), and therefore lends credibility to the view of the EU as a "regulatory state" (Majone 1997). Environmental regulation has its starting point in welfare economics and the study of market failure, and the need to compensate with regulations.

Our aim here, however, is not to try to explain the more general reason for regulation and in particular EU regulation in environment policy (or, more precisely, energy and climate change policies). Rather, we aim to explain the added value of regional intervention to secure this type of regulation, and the particular role played by the European Commission in combination with the German and French EU Presidencies in this regional policy-making process.

In this task, traditional theories of European integration and regional cooperation in general will be more suitable as a theoretical framework than regulation theories. In particular we argue that the multi-level governance represents an interesting alternative approach to the more state-centred approaches in liberal inter-governmentalism, when one tries to explain the role of different institutions (such as the European Commission) and the interplay between them (for example, the Commission and the various Presidencies).

Different variations of multi-level governance have been developed after Marks' first model in 1992 (Marks 1992; Marks et al. 1996; Hooghe and Marks 2001, Fairbrass and Jordan 2004). The central assumption in addition to the reduced influence of the nation state is the particular relevance of supranational institutions. The Commission, as one example, exercises an independent influence over the policy-making which exceeds their role as agents for national governments (Hooghe and Marks 2001:3). As Skjærseth and Wetterstad (2008) point out, this implies that individual governments do not have full control over collective decision-making at the European level in areas like environmental policy. This argument is in line with much reasoning in different new institutionalist approaches to European integration, and points to the independent role of the Commission and its ability to play on different institutional mechanisms embedded in the very construction of the European Union. The interplay between the Commission and the role of the EU Presidencies is an example of one such institutional mechanism (Tallberg 2006). The European Commission's development of several of the different elements in the climate change package, as well as the way the final agreement was reached among the member states and the different EU institutions are in line with this theoretical reasoning.

Climate policy in the EU involves a complex distribution of powers and responsibilities between the EU and the member states. The EU institutions can act only to the extent that they have been given the competence to do so by the member states of the Treaties establishing them. Whenever we consider "EU policy" in the field of climate change and energy, it is important to bear in mind that this term refers to a combination of policies and measures decided and implemented by the supranational institutions of the EU and by national (and, in some cases also sub-national) institutions in 27 member states.

The European Union is an umbrella concept and an institutional framework uniting all the various forms of cooperation under different Treaties between the member states. It has also become the political identity under which the member states act collectively on the international scene. This is particularly significant to bear in mind in this context because EU climate change policy originated as part of the Union's external environmental policy in the early 1990s.

One of the objectives of EU environmental policy, as laid down in Article 191(1) of the TFEU, is "promoting measures at an international level to deal with regional or worldwide environmental problems". To achieve this objective, the EU can adopt internal legislation, but also "cooperate with third countries and with the competent international organisations" by concluding international agreements. When multilateral negotiations on climate change started in the UN, the EU member states decided to participate in these negotiations as a single block on the basis of a common position. Thus the EU became one of the main actors in the global negotiations, even though, at the time the United Nations Framework Convention on Climate Change (UNFCCC) was signed, it had not yet adopted any internal legislation to deal with climate change. Its common position was based on political consensus between the member states and an aggregation of their emerging national policies. Gradually, these national policies were complemented and supported by "common and coordinated policies and measures" at the EU level, including a number of important legislative measures.

The EU institutions can adopt environmental legislation binding on all member states without their unanimous consent; a "qualified majority" of member states votes is sufficient, except in two cases relevant to climate change. Under Article 192(2) unanimity is still required for any "provisions primarily of a fiscal nature" as well as for "measures significantly affecting a Member State's choice between different energy sources and the general structure of its energy supply". The first exception was invoked in the 1990s to block a Commission proposal for a harmonized carbon/energy tax to be introduced throughout the EU as a climate policy measure. The second has never explicitly been invoked so far but is looming in the background in all political decision-making on climate change, especially as the impact of climate measures on energy policy is increasing.

8.2   EU CLIMATE CHANGE POLICIES IN A GLOBAL PERSPECTIVE

Climate change policies have developed mainly over the last two decades, and are therefore a relatively new policy area. In February 2010 the EC established a separate DG for climate action, DG CLIM (Press Release, 17 February 2010). This is a sign of the growing importance of climate change policies within the Community.

However, EU policies do not develop in a vacuum. International, national and EU politics influence each other in a multi-level fashion. The EU has been talking about initiatives to reduce CO_2-emissions since the early 1990s, then with special focus on emissions from transport and energy efficiency. The policies remained vague and were repeatedly postponed and weakened (Transport and Environment 2009).

A new impetus appeared in the late 1990s, when the Community started its search for how best to reach its Kyoto targets and committed itself to a global leadership role in the combat of climate change. Under the Kyoto Protocol[1] from 1997 the EU committed to an 8% reduction by 2010 compared to 1990 emissions. The international climate regime offered the EU the chance of showing international leadership, particularly starting in 2002 when it became clear that the Bush administration was not going to ratify the protocol because of is perceived costs to the American economy.

The Protocol entered into force in February 2005. After the US decided to withdraw from Kyoto, the EU used a lot of effort to hold other countries (especially Russia) to their word and succeeded in getting enough countries to sign so the treaty could enter into force. However, the fight against climate change competed for attention with the EU's own Lisbon priorities of competitiveness, jobs and economic growth (and subsequently also with the financial crisis that started in 2007).

The 2008 package was composed of a range of elements, most of which were not new to the EU climate policy agenda. Nevertheless, their combination into a package has been argued to present a "take it or leave it" approach (Helm 2009) that made its adoption feasible. For example, despite many parliamentarians' complaints that the changes to the ETS were too generous to industry, and that the possible use of the Clean Development Mechanism (CDM) allowed member states to undertake most of their emissions reductions outside of Europe, they still voted in favour. The feeling was that voting against it would leave the EU without a climate policy, and that what was being agreed was – despite its shortcomings – a better option.

---

1   The Kyoto protocol from 1997 determined binding targets for GHG emissions reductions under the UN Framework Convention for Climate Change, the UNFCCC.

On a regional level, the EU started considering the respective role of "common and coordinated" versus national policies and measures as a means of fulfilling its collective quantified emissions reduction target of 8%. This debate involved conflicting interpretations of the principle of subsidiarity.[2] Some member states argued that national measures would be sufficient to reach their targets, while others considered necessary a range of harmonized measures at the EU level. In June 1998, the EU Council reached political agreement on the principle of internal "burden-sharing" – i.e. the alloca-tion of responsibility to individual member states for the achievement of the common Kyoto target – as well as on the need for further development of common measures.

The EU's commitment under the Kyoto Protocol spurred the first *European Climate Change Programme* (ECCP) (2001–2003). It was launched by the Commission in June 2000, with the goal to identify and develop all the most environmentally beneficial and cost-effective additional policies and measures enabling the EU to meet its Kyoto target (CANEurope et al. 2005). It was a stakeholder structure under which the Commission debated with industries and NGOs and prepared new cost-effective measures to fight climate change. Eleven working groups were established.[3] The first ECCP identified more than 40 measures that potentially could cut emissions by twice the required level (DG Environment, 2002). This shows that the biggest challenge was not technological (assuming that the calculations were right), but political.

The second ECCP was launched in 2005 to "explore further cost-effec-tive options for reducing greenhouse gas emissions in synergy with the EU's 'Lisbon strategy' for increasing economic growth and job creation" (DG Environment 2006). In a background note the Commission set the goals for a second ECCP: energy efficiency, renewable energy, the transport sector (including aviation and maritime transport), and carbon capture and storage (European Commission 2005).

The Commission appointed a High Level Group on competitiveness, energy and the environment in February 2006. It functioned as an advisory plat-form bringing together all relevant stakeholders, including several Members of the Commission. Several of their recommendations were incorporated

---

2   The subsidiarity principle as laid down in Art. 5 of the Treaty provides for common action to be taken "only if and so far as the objectives of the proposed action cannot be sufficiently achieved by the Member States and can, therefore, by reason of the scale or effects of the proposed action, be better achieved by the Community".

3   The 11 groups covered emissions trading; JI/CDM; energy supply; energy demand; energy effi-ciency; transport; industry; research; agriculture; sinks in agricultural soils; forest-related sinks.

into the Commission's Action Plan on energy efficiency from October 2006 (European Commission 2006). *The multiple goals of* security of energy supply, reduction of carbon emissions, fostering competitiveness and technological development were used to advocate the need for increased energy efficiency.

The ECCP is a part of the learning and legitimising process of EU decision-making. It is important because it brings together actors from a wide variety of fields, including the EU bureaucracy, member states, NGOs, and industry. However, it is not a formal structure for any of the EU decision-making on climate policies. It remains a programme wherein policy reviews are being undertaken, as well as important policy learning and development, and, crucially, legitimacy building.

Europe is making good progress towards meeting targets set by the Kyoto Protocol in its second phase, which runs from 2013 to 2020. In 2012 combined emissions from the 28 EU member countries were estimated at 18% below the 1990 level, compared to the EU's commitment to reduce its GHG emissions by 20% below base-year levels.

However, since Russia, Japan and New Zealand have decided against taking part, Canada has withdrawn from Kyoto entirely and the developing countries are not required to limit their emissions at all, the second Kyoto period affects only around 14% of global emissions. Therefore, on the initiative of the EU, the UN climate change conference in Durban, South Africa in 2011, decided to launch the negotiations on a new global climate agreement applicable to all countries, developed and developing alike. The new agreement is to be adopted in 2015 and to enter into force in 2020 (European Commission 2013).

8.3   THE COMPOSITION OF THE PACKAGE AND THE ROLE OF THE EUROPEAN COMMISSION

We now turn to the issue of how the European Commission in its different proposals structured the total package and worked with the member states regarding the impact of policies on individual states. What was the Commission's strategy, and how was this influenced by the international climate regime? How did the Commission accomplish the task of creating a total package acceptable for all the different member states and the very low-carbon ambitious European Parliament?

We find it significant that in general, the European Council gave the Commission the control over the time schedule and the preparation of both the documents and the detailed communications with the member states about the assessments of the impact on their country. This implies that the Council viewed the Commission as the most effective tool to achieve an EU-wide agreement, most possibly building on its extensive consultation efforts of the previous half decade.

It was also significant that the Commission had managed to achieve general acceptance for the EU ETS which in January 2005 commenced operation as the largest multi-country, multi-sector Greenhouse Gas Emission Trading System worldwide. The system is based on the idea that a price on carbon emissions will internalize the cost of environmental damage (MacKenzie, Hanley, and Kornienko 2008). It will furthermore ensure that the most cost-effective changes will take place first, and that it will increase investment in low-carbon alternative energy technologies, because they are expected to become profitable. This approach satisfied the market-liberal countries in the EU, which approved the market-based mechanism. It furthermore satisfied the new member states, which benefited from the allocations system and saw the scheme as potential foreign investment in the country (Fazekas, 2008).

The Commission's strategy focused on two important issues: on the one hand, linking energy policies and climate change policies, and on the other hand, addressing individual member states' concerns about cost implications. Many countries, maybe particularly those in Eastern Europe, were concerned about having to take on large commitments while also strengthening their weak economies.

The European Commission presented the energy and climate change package in January 2007, and was the key actor in developing and implementing a strategy which led to the adoption by the Council of Ministers and the European Parliament in December 2008. The package was aimed at multiple goals: combating climate change, ensuring energy security, addressing the issue of renewable energy and creating a real market for energy. The Commission's proposal was built around three pillars:

1. A true internal energy market, meaning real choice for consumers with further liberalization;
2. On the energy supply side, accelerating the shift to low-carbon energy. 20% of the energy mix should derive from renewable energy sources, including a minimum of 10% of transport petrol and diesel from biofuels. In order to accelerate the development and deployment of cost-effective low carbon technologies, a Strategic Energy Technology Plan (SET Plan) was set up by the Commission;
3. On the energy demand side, increase energy efficiency, particularly in transport and domestic energy use.

The package was deliberately composed as an *energy and climate change* package to reach both important energy and climate goals for the Union and its member states. For many countries the tangible need for energy security was a stronger driver than concerns for the effects of climate change. By combining the two issues, the Commission effectively framed climate change policies in terms that were more relevant to the political situation, particularly for the new member states in Eastern Europe[4] (personal interviews). Several interviewees who had been following the negotiations closely expressed the view that the energy security issue, which recently had been highlighted by the conflict between Russia and Ukraine over delivery of Russian gas to Europe through Ukraine, was from the outset seen as the obvious starting point for achieving acceptance for climate change policies in the new member states. By selling reduction of GHG emissions as a consequence of policies aimed at decreasing the dependence on foreign energy resources, the Commission made it tempting for initially reluctant member states (whether they thought it would be too costly or too technologically challenging) to accept the package. The same argument was also used to increase the degree of liberalization in the energy market.

The second oft-mentioned element in the Commission's strategy was the detailed calculation of the costs of the different elements in the package for all the member states. The Commission simultaneously made several efforts to explain the calculations and to sell the idea to the different member states that these figures were reliable and not as costly as many of the member states had feared. Commission representatives were travelling to all the different capitals to show and discuss the calculations with the national governments (personal interviews).

This argument was also linked to the introduction of burden-sharing among the member states. At the occasion of the Council summit in Germany in 2007, the Commission was called to propose a system for burden-sharing among the members which was accepted and seen as very important for the final outcome.

The use and further development of ETS (see part 5) was also important in the construction of the total package, but several of the main actors also felt the need for more regulation in addition to the market mechanism introduced in the ETS scheme.

At the summit in Paris in December 2008, EU leaders reached agreement on the package after informal negotiations between the Parliament, the Commission and the French Presidency. The Parliament finally adopted it on

---

4   The only important energy issue left out was nuclear energy, which was left to the individual member states to decide upon. The issue is highly contentious with member states holding strongly opposing views (e.g., France producing much of its electricity from nuclear plants, while Denmark is highly opposed, and Sweden and Finland having decided to phase out nuclear energy).

17 December 2008. The aim was to deliver the bloc's ambitious objectives of slashing greenhouse-gas emissions, boosting renewable energies by 20% by 2020 and reducing the Union's dependency on imported fuels, as well as enabling the EU to halve emissions relative to 1990 levels by 2050.

The progress toward the 2020 targets is noticeable already as the results achieved by 2012 show 18% reduction in GHG, an increase in the share of renewables to 14.4%, and 8% reduction in primary energy consumption between the 2006 peak and 2012. Beyond the short-term impact of the crisis, the EU is steadily decoupling growth in economic activities and GHG (European Commission 2014).

Expanding the 2020 strategy, in 2012 the European Commission adopted a legislation setting non-binding national targets for improving energy efficiency, launched the European Climate Adaptation Platform as a part of EU Adaptation Strategy, and it launched a pan-European communication campaign with the slogan "A world you like, With climate you like". In March 2014 the Commission presented the 2030 policy framework, which seeks to drive continued progress towards a low-carbon economy. However, the EU leaders have not yet decided on the framework (European Commission 2014).

8.4 EMISSION TRADING AS THE ROAD TO SUCCESS?

The EU ETS has been hailed as the cornerstone of EU climate policy. The idea of an EU-internal pilot trading scheme had been presented already in 1998 in a communication from the Commission on an EU-post 2012 strategy, but was taken considerably further in the Green Paper on ETS from March 2000 (Skjærseth and Wettestad 2009:107). The Commission crafted agreement among central actors and stakeholders through their European Climate Change Programme, whose working group on emissions trading stressed in their conclusion in June 2001 the need to implement emissions trading as soon as possible.

Skjærseth and Wettestad (2009; see also Convery 2008) show that the Commission played a major role in changing the EU stance on international emission trading after the adoption of the Kyoto Protocol in 1997. They point to three salient explanatory factors. Firstly, that the commitments made in the Kyoto Protocol were perceived as mandatory by the Commission and the member states. Secondly, there was an important change of personnel in the climate change unit at the DG Environment in 1998, and the new team was made up of economists who were perceptive to the advantages of emissions trading as a useful tool to reach the Kyoto targets. Thirdly, emissions trading would be subject to majority voting, thus avoiding the fate of the carbon tax proposal, which required unanimity and would not be adopted.

The EU ETS started operating in 2005, and ran for two years as a trial period before the Kyoto period 2008–2012. It has become a generally accepted mechanism throughout the Union, but its implementation has been the source of several divisive issues. Firstly, there was the issue of the overall number of credits in the system. At the end of the trial period, pollution credits were shown to be grossly over-allocated by several countries, forcing down carbon prices and undermining the scheme's credibility. This prompted a reduction of allocations in 2007 (Euractiv, 2010a). Secondly, there was an issue with how these credits (of which there had proven to be too many) should be allocated to installations within member states (the issue of national allocation plans – NAPs). Thirdly, there was the question of whether these credits should be paid for by industry or given for free. Given these problems in January 2008 the European Commission proposed revising the system. The derogations to the revised ETS constituted one of the main changes to the Commission's proposal during the negotiations. We shall here look briefly at the three issues outlined.

First, the early *over-allocation of credits* resulted mainly from the fact that the initial amount of credits was determined on the basis of governments proposing to the Commission the amount of credits allocated to their country. The governments generally asked their domestic industries, which had an obvious incentive to overstate their emissions. The governments, in turn, had the same sort of incentives relative to the Commission. It led to a situation in which some industries received more permits than they needed to cover their total emissions. When this problem of over-allocation became known in 2007 the price of permits collapsed (Buchan 2009).

Secondly, the *national allocation of permits* was not specified in the EU ETS, but was left to member states' national allocation plans (NAPs). The NAPs, however, had to be submitted to the Commission for approval. For various reasons, most member states were not able to finalize their NAPs before the December 2007 deadline, which subsequently hampered their ability to comply with the March 2008 deadline for finalized emission accounts. This created a legal conflict between member states and the Commission (Peeters 2008).

Thirdly, the question of *free allocations vs. auctioning* was another source of tension. Free allocation of permits is an important distortion of the market, but was introduced to give industry time to readjust and to prevent "carbon leakage" (industry relocating to areas with less or no charge on GHG emissions). Free allocations have in some cases given rise to important windfall profits to companies that have been able to charge customers for carbon permits that the companies in the end were given for free. Moreover, some of the new member states in Eastern Europe received permits to cover emissions for the whole production or generation of new plants, which reduced their incentive to invest in low-carbon industries (Buchan 2009).

The Commission's original proposal was for 100% auctioning to the power sector in phase III (from 2013). Poland (generating over 80% of its power from coal-fired plants) called into question auctioning for the power sector and, with the support of other new member states, succeeded in securing exemptions for its power sector – this would result in auctioning gradually being phased-in, rising from an initial 30% to full auctioning in 2020. For non-power sectors installations, it was proposed that 20% of permits be auctioned from 2013, rising to 100% in 2020 (Stop Climate Change 2008).

In order to ease the energy transition for countries with high dependence on fossil fuel or insufficient connection to the European electricity network, it was negotiated that ten member states can apply for reduced auctioning rates in power production: at least 30% in 2013, gradually rising to 100% in 2020 (Press release Council, 6.4.09). "This concession was tailored to new member states by stating that the phased auctioning option was open to states ill-connected to the continental European grid (such as the Baltic states) or states at least 30% dependent on a single fossil fuel (coal in Poland, gas in Hungary) or states with income per head of only half the average (Balkans)" (Buchan 2009: 126). The larger amount of permits given to "less wealthy" (mainly Eastern European) states was presented as a "solidarity mechanism", because the states will get the opportunity of having substantial incomes from selling allowances (Press release, Council, 6.4.09).

The outcome of the negotiations was that 20% of permits will be auctioned from 2013, gradually increasing to 70% in 2020, with a view to reaching 100% in 2027 (Press release, Council, 6.4.09). The auctioning method seems to require lower government costs because it leaves the distribution of the allowances to the discretion of industries. Industries will buy the allocations they need, taking into account their cost of abatement of the GHG (Peeters 2008).

It seems clear that the EU ETS provided a (perceived) cost-effective mechanism for the EU to reach its Kyoto target. Cost-effectiveness is politically easier to sell than direct regulation and "red tape", and the ETS was therefore clearly important to the success of the climate change package. However, in addition to its market-based foundations, which resonated well with a growing number of economists in public administrations, the ETS was able to provide a practical solution to the difficult question of effort-sharing between member states, and even in the end a "solidarity mechanism" through its capacity for economic transfers between countries.

In January 2013 a major reform took effect, which included the introduction of a single EU-wide cap on emissions, which will be reduced by 1.74% each year so that by 2020 emissions will be 21% below the 2005 level. The other major change was that the free allocation of allowances is now progressively being replaced by auctioning, starting with the power sector (European Commission, 2014).

8.5   THE ROLE OF THE PRESIDENCIES: THE GERMAN AND THE FRENCH PRESIDENCIES AND THE END GAME

As we have shown, most of the elements in the 2006 proposal document from the Commission had been developed through extensive consultation processes (in particular under the auspices of the European Climate Change Programme) over the previous six years. However, although the EU-15 was on track to meet its Kyoto targets, the situation was less optimistic for EU-27 (EEA 2009).

Several of our interviewees (most of whom were key decision-makers) emphasized the role of the presidencies and the interplay between the Commission and the different presidencies. We shall therefore use this section to look more closely at what the German and the French EU Presidencies did and try to shed light on the complex multi-level interactions between institutions and actors. To what extent were these countries engaged in "agenda management, brokerage and representation" (Tallberg, 2006: 82)?

The ambition for the EU to be a global leader in climate policy terms, and particularly for the Union to play a decisive role during the Copenhagen summit, transpires as a major motivation for both the German and the French Presidencies. According to the German Presidency, 2007 was marked by the need for "redressing the internal and the external 'credibility gap' in the energy and climate change policies of the EU" (Oberthur 2008). Angela Merkel stated that "It is important we can tell the G8 members that Europe has made a real commitment. That gives us a measure of credibility" (German Presidency Press Release, 09.03.2007). The German Presidency wanted to propel a "pioneering role at the global level in combating climate change" (German Presidency Programme 2007) and assured that "climate change will remain a high profile priority for the Union also during the first Trio-Presidency" (18-month programme with Portugal and Slovenia). In order to achieve this, "Europe has to opt for a strategic partnership with Russia. A transatlantic alliance is also important" (A. Merkel, speech at the 50th anniversary of the Rome Treaties). Also the French government had high ambitions for a European climate and energy policy during it Presidency. The fight against climate change was at the top of its professed agenda for the

Presidency period, with targets for emissions reductions and for renewable energies given special mention (Fillon 18.6.08).

The idea of lifting the issue up from the day-to-day process of policy-making to become a high-politics issue emerged under the German Presidency. Interviewees acknowledged that the German Presidency was central in propelling the European Commission to investigate how 20-20-20 could be achieved. The Presidency managed to ensure that during the March 2007 summit, the binding target of 20% reduction in GHG emissions by 2020 was adopted, with the possibility of a binding target of 30% if other developed countries would take on similar obligations. The March 2007 summit also established that the EU aimed for a 20% reduction in energy use through energy efficiency measures by 2020 compared to the BAU (business as usual) projections, and it endorsed a binding target of 20% renewable energy in energy consumption by 2020, including 10% biofuels for transport (Presidency conclusions, European Council, 8/9 March 2007).

Angela Merkel contributed significantly to the brokering of a deal on renewables, particularly with the Czech Republic, Poland and France. Having established the overarching goals, the ground was prepared for negotiations between member states on key issues. These included a review of the ETS for the period beyond 2012, how the responsibility for reaching the 20% reduction target should be distributed between states, a new framework for the promotion and trade of renewable energies, including biofuels, a mechanism to finance 12 CCS demonstration plants by 2015 and a legal framework for CO_2 storage. The Council therefore "invited" the Commission "immediately to start a technical analysis of criteria, including socio-economic parameters and other relevant and comparable parameters, to form the basis for further in-depth discussion" (European Council March 2007: 12). The Council also asked the Commission to undertake a review of the ETS (ibid., p. 13).

Both the German Presidency and the Commission experienced internal divisions, particularly with regard to the issue of vehicle emissions. Transport being one of the largest sources of GHG emissions, the environment Commissioner Dimas had proposed to legislate for a maximum of 120 g CO_2/km emissions from new cars from 2012. The German car industry was heavily opposed to such legislation, and demanded the plans withdrawn. The German Minister of the Environment Gabriel (then in the Presidency) sided with Dimas, whereas both the Economy Minister Glos (in the Presidency) and Commissioner Verheugen (in the Commission) supported the industry (EU Observer 2007). In the end the Commission proposed to legislate for average emissions from new cars to 130 g/km by 2012, allowing for tyre makers, fuel suppliers and others to ensure another 10 g/km of emissions reductions so the overall object was 120 g/km for new cars by 2012.

Some scholars (e.g., Lefebyre, 2009) point to the voluntarism and pragmatism of Sarkozy in managing the final deal of the climate change package. Others point to the role of the French ambassador to the EU in coordinating the decision and the summit with the European Parliament (personal interviews). However, the establishment of a broad policy network of entrepreneurs was important to the December summit's success (Braun 2009).

The French Presidency had to deal with several difficult issues regarding the climate change package. The Eastern European states were apprehensive about having too costly measures forced upon them, particularly because if the package was subject to conventional environmental policy rules, a majority decision would be sufficient. Furthermore, Poland, Germany and Italy all threatened to oppose any deal because of the perceived cost of auctioning emission permits under the ETS from 2013 in the face of the economic recession (Buchan, 2009). In addition, the Commission's proposal on vehicle emissions exposed strong disagreement between Germany and France – the German car industry wanting exemptions for heavier and powerful cars, the French wanting full implementation.

The issue of costs particularly for the Eastern European states was solved through a deal between President Sarkozy and the Polish Prime Minister Donald Tusk that any overall deal would be reached by unanimity. The issue over auctioning ETS permits, as we saw in the previous section, resulted in a derogation from the Commission's overall proposal. Under the deal, full auctioning of carbon allowances may take place as late as 2027, instead of 2020 as initially proposed. Moreover, industry sectors threatened by third-country competition were promised a continuation of free allowances. In a move considered crucial for the deployment of "clean coal" technology, 300 million ETS allowances were secured to fund carbon capture and storage (CCS) projects in the EU.

When it came to vehicle emissions, a compromise between President Sarkozy and Chancellor Merkel paved the way for an agreement at the end of 2008 on 120g/km emissions from new cars by 2015, and phasing in penalties on car makers for exceeding this limit (Buchan, 2009). This compromise, made by the "oddest of couples" indicates the huge importance of an agreement on a climate change package for the two large countries who had invested much political prestige in the project to place "Europe in the vanguard of the fight against climate change and reinforcing energy security" (French Presidency Programme June 2008).

CONCLUSION – EU PROCESSES AS POLITICAL ENGINEERING

In this paper we have analysed how and why the EU was so successful in developing and adopting the climate change package in December 2008. In particular we have looked at the extent to which this success could be attributed to the role of the European Commission. We got strong indications that the ability of the EU actors, in particular the Commission, to employ an interplay between several different decision-making strategies and instruments of supranational influence in relation to the member states was the most important factor explaining the success. There seem to be three main elements contributing to this.

Firstly, the deliberate combination from the outset of energy and climate change strategies. This idea seems to have been developed by the Commission, but supported by key member states, among them France and Germany. Actions to increase energy security for member states were combined with actions to reduce climate change – the Commission argued that reduction of fossil fuel consumption, energy conservation and increased supply of renewable energies will reduce the dependency on foreign import, as well as lowering GHG emissions. Thus, for several countries, especially new member states, this was mainly an energy package whereas for most of the old member states and the Commission it was mainly a climate change package.

Secondly, several of our interviewees in the Commission argue that the Commission strategy to calculate the real costs for different proposals for all the member states to show both that it was not too costly for them and that the burdens were shared in a balanced manner between the member states was also important. This transparency and realism had also shown that the overall costs for each country were manageable. This contributed to the overall acceptance of the ambiguous goals. This was important both during the negotiations between the member states and the EC in 2007 and 2008, and in December 2008 when France was able to make an agreement on the package at the summit in Paris and also with the European Parliament afterwards.

Thirdly, we highlighted the role of the Presidencies in interplay with the Commission and the ability of the Commission to exploit the German and French Presidencies, first to set the 20-20-20 goals (Germany 2007), and then to conclude the deal (France 2008). This ability to shift the issue from low politics in the detailed proposals from the EC to high politics using the summits to make the decisions on the principles and then take the proposals down again, the normal logic of EU decision-making was a key element in managing the differences between the member states.

In addition, we see the existence of the ETS, which established a market-based mechanism to deal in reductions and to implement the decisions as important. This gave national (particularly UK) politicians better chances of selling the package at home, and proved to be an instrument which could be given a host of different intentions and effects (e.g., cost-effectiveness, effort-sharing, solidarity). But again, the ability of the EU to combine these kinds of market mechanisms with regulations seems to be as valuable for a package solution.

These elements could be seen as an argument for the explanatory power of the multi-level governance theory for the case we are studying. There are elements in the EU construction which, independent of the member states, can be used by the EC to exercise influence over the outcome of specific decisions by combining issues, using the role of the EC in preparing summit decisions and the role of the EU Presidencies in orchestrating a decision-making logic partly outside the normal procedures in the Treaty.

The final decision on the climate change package was made more than 4 years ago. The pace of the decision-making process was not influenced by the financial crises. The implementation of the package in the member states has now started. We don't know if the long term effects of the financial crises in many West and East European countries and the problems in the euro zone countries will influence the implementation, but our aim in this paper was to look more into the development and the final decision-making process of the package. Seen from this perspective the additional capacities of a regional organization have proven to be of importance for the success of the efforts made to strike a deal at the regional level in the difficult climate change area.

BIBLIOGRAPHY

Beunderman, M. (2007). German EU presidency struggles with climate change, *EU Observer*, http://euobserver.com/9/23365, last accessed on 03.04.2010

Braun, M. (2009). The evolution of emissions trading in the European Union – The role of policy networks, knowledge and policy entrepreneurs, in *Accounting organizations and society, 34*(3-4), 469–487.

Buchan, D. (2009). *Energy and Climate Change, Europe at the crossroads*, Oxford: Oxford University Press.

CANEurope, FOE Europe, Greenpeace, T&E, WWF (2005). *Input from environmental NGOs at the start of the next round of the European Climate Change Program (ECCP)*.

Convery, F. (2009). Origins and Development of the EU ETS, in *Environmental and Resource Economics, 43*(3), 391–412.

Council of the European Union (2007). *Trio programme*, http://www.
eu2007.de/includes/Download_Dokumente/Trio-Programm/
trioenglish.pdf, last accessed on 03.04.2010.

Council of the European Union (2009). Council adopts climate-energy
legislative package, *Press Release*, http://www.consilium.europa.eu/
uedocs/cms_data/docs/pressdata/en/misc/107136.pdf, last accessed on
03.04.2010.

Euractiv (2010a). EU Emissions Trading Scheme, Dossier, http://www.
euractiv.com/en/climate-change/eu-emissions-trading-scheme/
article-133629, last accessed on 03.04.2010.

Euractiv (2010b). *Brussels readies for tough climate negotiations*, http://www.
euractiv.com/en/climate-change/brussels-readies-tough-climate-
negotiations/article-176457, last accessed on 03.04.2010.

European Commission (2002). EU focus on climate change, *DG
Environment*, http://ec.europa.eu/environment/climat/pdf/climate_
focus_en.pdf, last accessed on 03.04.2010.

European Commission (2006). The European Climate Change Package.
EU Action against Climate Change, *DG Environment*, http://ec.europa.
eu/environment/climat/pdf/eu_climate_change_progr.pdf, last accessed
on 03.04.2010.

European Commission (2010). Commission creates two new
Directorates-General for Energy and Climate Action, *Press Release*,
17 February 2010, http://europa.eu/rapid/pressReleasesAction.
do?reference=IP/10/164&format=HTML&aged=
0&language=EN&guiLanguage=en, last accessed on 03.04.2010.

European Commission (2013). *The European Union explained: Climate action*,
http://europa.eu/pol/pdf/flipbook/en/climate_action_en.pdf

European Commission (2013). *The EU Emission Trading System (EU ETS)*,
http://ec.europa.eu/clima/publications/docs/factsheet_ets_en.pdf

European Commission (2014). *The 2020 climate and energy package* [online]
http://ec.europa.eu/clima/policies/package/index_en.htm

European Commission COM (2005). *35 final "Winning the Battle Against
Global Climate Change"*, http://eur-lex.europa.eu/LexUriServ/
LexUriServ.do?uri=COM:2005:0035:FIN:EN:PDF, last accessed on
03.04.2010.

European Commission COM (2006). *545 final "Action Plan for Energy
Efficiency: Realising the Potential"*, http://eur-lex.europa.eu/LexUriServ/
LexUriServ.do?uri=COM:2006:0545:FIN:EN:PDF, last accessed on
03.04.2010.

European Council (2007). *Presidency conclusions*, 8/9 March 2007, http://europa.eu/european-council/index_en.htm, last accessed on 03.04.2010.

European Federation for Transport and Environment (2009). *Reducing CO₂ Emissions from New Cars: A Study of Major Car Manufacturers' Progress in 2008*, Brussels.

Fairbrass, J. and Jordan, A. (2004). European Union Environmental Policy: A Case of Multi-level Governance? in M. Flinders and I. Bache (Eds.) *Themes and Issues in Multi-Level Governance*, pp. 147–164. Oxford: Oxford University Press.

Fazekas, D. (2008). *Hungarian Experiences with the EU ETS*. http://www.aprec.info/documents/08-08-hungarian_experience_with_the_eu-ts.pdf, last accessed on 03.04.2010.

Fillon, F. (2008). *Discours du premier ministre lors du débat portant sur la Présidence française de l'Union européenne*, 18.6.08, http://www.gouvernement.fr/premier-ministre/discours-du-premier-ministre-lors-du-debat-portant-sur-la-presidence-francaise-de-l, last accessed on 03.04.2010.

French Presidency of the Council of the European Union (2008). *Work Programme*, http://www.france-science.org/spip.php?article964, last accessed on 03.04.2010.

German Presidency (2007). Historical agreement on climate protection, *Press Release*, 09.03.2007, http://www.eu2007.de/en/News/Press_Releases/March/0309BKBruessel.html, last accessed on 03.04.2010.

German Presidency of the Council of the European Union (2007). *Work Programme*, http://www.eu2007.de/en/The_Council_Presidency/Priorities_Programmes/index.html, last accessed on 03.04.2010.

Helm, Dieter (2009). EU Climate Change Policy – A Critique, in D. Helm and C. Hepburn (Eds.), *The Economics and Politics of Climate Change*, pp. 222–244. Oxford: Oxford University Press.

Hooghe, L. and Marks, G. (Eds.) (2001). *Multi-level governance and European integration*, Lanham, MD: Rowman & Littlefield Publishers.

Lefebvre, M. (2009). *An Evaluation of the French EU Presidency (ARI)*, Elcano Royal Institute, http://www.realinstitutoelcano.org/wps/portal/rielcano_eng/Content?WCM_GLOBAL_CONTEXT=/elcano/elcano_in/zonas_in/europe/ari43-2009, last accessed on 03.04.2010.

MacKenzie, I., Hanley, N. and Kornienko, T. (2008). The Optimal Initial Allocation of Pollution Permits: A Relative Performance Approach , in *Environmental and Resource Economics*, *39*(3), 265–282.

Majone, G. (1997). From the Positive to the Regulatory State. Causes and Consequences of Changes in the Mode of Governance, *Journal of Public Policy*, 17(2), 139–167.

Marks, G. (1992). Structural Policy in the European Community, in A. Sbragia, (ed.), *Europolitics: Institutions and Policy Making in the 'New' European Community*, pp. 191–224. Washington D.C.: The Brookings Institution.

Marks, G., Scharpf, F., Schmitter, P. and Streeck, W. (1996). *Governance in the European Union*, London: Sage.

Merkel, A. (2007). *Speech at the 50th anniversary of the Rome Treaties*, http://www.eu2007.de/en/News/Speeches_Interviews/March/0325BKBerliner.html, last accessed on 03.04.2010.

Oberthür, S. (2009). The role of the European Union in global environmental and climate governance, in M. Telò (Ed.) *The European Union and global governance*, pp. 192–210. London: Routledge.

Peeters, Marjan (2009). Legislative choices and legal values: considerations on further design of the European greenhouse gas ETS from a viewpoint of democratic accountability, in M. Peeters and M. Faure (Eds.), *Climate Change and European Emissions Trading: Lessons for Theory and Practice*, Cheltenham: Edward Elgar Publisher.

Princen, S. (2009). *Agenda-Setting in the European Union*. London: Palgrave Macmillan.

Skjærseth, J.B. and Wettestad, J. (2008). Implementing EU emissions trading: success or failure?, in *International Environmental Agreements: Politics, Law and Economics*, 8(3), 275–290.

Skjærseth, J.B. and Wettestad. J. (2009). The Origin, Evolution and Consequences of the EU Emissions Trading System, in *Global Environmental Politics*, 9(2), 101–123.

Stop Climate Change (2008). *EU emissions trading scheme final negotiation outcome*, http://www.stopclimatechange.net/index.php?id=68, last accessed on 03.04.2010.

Tallberg, J. (2007). *Bargaining power in the European Council*, Stockholm: Swedish studies for European Policy Studies.

Part III Climate Policy Actors and History

The third part reviews climate policy history and other major societal challenges, explaining why it has been so difficult to act resolutely. In particular the section discusses the ambitions and shortcomings of Norway in trying to be a role model country, something Randers has tried to promote since leading the governmental commission on Norway as a low emission society in 2006.

In Chapter 9 Marit Sjøvaag-Marino looks at the implementation of climate strategy in Norway. One of her observations is that Norwegian climate policy is continually drawn between lofty intentions and *realpolitik*, between the need for jobs and the side effects of an active fossil energy sector. The Norwegian petroleum industry must definitely be considered one of the polluters; however, the sector's importance for Norwegian wealth and job creation has made it difficult to impose policies that would seriously threaten its existence. Norway's lead climate negotiator for many years, Hanne Bjurstrøm, then gives an insider's view of Norway's role in the global climate negotiations. She argues that Norway has been given a fairly influential part in the negotiation process, "relative to our size and share of the total global emissions." She emphasizes that Norway was one of the first countries in the world to adopt a CO2 tax (1991) and to implement green tax reforms. Gudmund Hernes (Chapter 10) concludes the section by discussing the shifting cycles of "Optimism and Pessimism in Climate Policy" over previous decades.

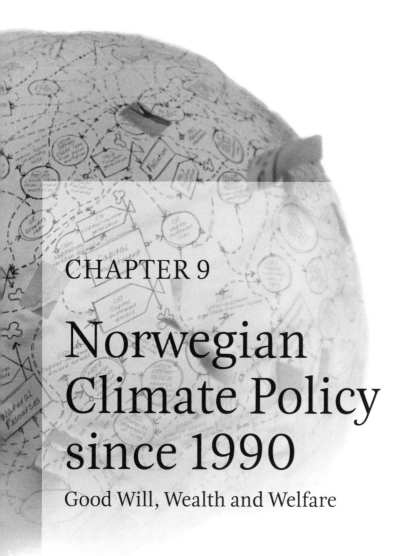

CHAPTER 9

Norwegian Climate Policy since 1990

Good Will, Wealth and Welfare

By Marit Sjøvaag Marino

INTRODUCTION

If we are to understand Norwegian climate policy, it is worth remembering some facts: Norway is a small country with an open economy, which has become rich particularly from exporting oil and gas. A substantial part of national energy consumption comes from hydropower. Atmospheric concentration of greenhouse gases is a global issue. The scientific consensus is that climate change is caused by human activity, of which use of fossil fuels is the main driver, but where changes in land use and deforestation also are important contributors.

Norway often claims to be, and is seen as, an environmentally friendly country. It was the first in the world to establish a separate ministry for the environment already in 1972. As shown in the chapter by Bjurstrøm in this volume, it has been an important contributor to international negotiations with the aim of establishing a global regime and accords for emission reductions. Moreover, it has in recent years been one of very few countries to follow up its pledges to support poorer countries with real money as well as administrative capacity, through the Green Fund, the UN-REDD programme, and its own Climate and Forest Initiative.

However, in contrast to the situation in many EU countries and in the EU as a whole, Norwegian emissions have shown no significant decrease over the last two decades. Current emission levels are in the area of 4–5 per cent *higher* than they were in 1990, the year typically used as base year for emission target pledges. EU emissions have in the same period been reduced by 15–20 per cent.[1] Since the turn of the century numerous studies and reports have provided us with a long list of potential measures to ensure emissions reductions in the region of 50–80 per cent (compared to 1990 levels) by 2050, which is the magnitude of the challenge according to the IPCC.[2]

This chapter will give an overview of central elements and developments in Norwegian Climate Policy since 1990, predominantly focusing on inland policies and emissions, and thus largely excluding policies such as the climate and forest initiative. It will, however, include references to domestic energy policy, because this policy field is intrinsically linked to climate policy. Energy policy is also the proverbial "elephant in the room" in Norwegian climate policy debates.

---

1   The numbers vary between EU-15 and EU-28, but the downward trend is clear for both sets of countries (EEA 2014).

2   These numbers are set on the basis of a 50 per cent chance of not increasing global mean temperature by more than 2°C relative to pre-industrial times.

All historical overviews of policy developments must choose a point of departure. Sometimes events produce a natural starting point for history, but this is the exception rather than the rule. Most policy events are reactions to what went before, rather than profoundly new. This chapter starts in 1989, when Norway set a target for future CO_2 emissions. One could easily have gone back to 1986 and the publication of the Brundtland report, or to 1972 and the establishment of the Ministry of Environment. However, by covering the history from the 1990s onwards the chapter covers, in this author's view, the most relevant aspects to help understand where Norwegian climate policy has met the most difficult conflicts. This should help in understanding why we are where we are today, and where the national climate policy may be heading.

The chapter is organized by themes, and two issues are given special attention. Firstly, the question of direct regulation vs. economic incentives, and secondly, how much of Norway's emission reductions should be done nationally as opposed to abroad. These issues are pivotal in the narrative of Norwegian climate policy development. They constitute the intellectual bridge between what is being implemented, what measures are not adopted, and how public discourse shapes our conception of the possible as well as the common sense.

9.1   THE 1990S – FROM EMISSION TARGET
BY TAXATION TO INTERNATIONAL TRADING

In 1989 Norway became the first country to set a target for future CO_2 emissions, when the Parliament decided that Norwegian CO_2 emissions should be stabilized at the 1989 level within year 2000. There was broad political consensus that a stabilization target was the best way to ensure GHG abatement (Hovden and Lindseth 2002:146). The approach tied in well with other developments at the time. Norway had signed international agreements on emission reductions in air polluting substances such as SOx, NOx and CFCs, and the success of these agreements depended on all signatories meeting their obligations. In the case of CFCs, most commonly associated with the problem of the depletion of the ozone layer, two factors played a significant role in making an international agreement possible. Firstly, the number of industries using CFCs was relatively limited, thus reducing the number of interested parties. Secondly, commercial alternatives to CFCs existed. Both these elements worked in favour of reaching and implementing an international accord (Sprinz and Vaahtoranta 1994).

The adoption of a host of international agreements[3] induced a feeling of optimism regarding the international community's ability and willingness to come together to solve environmental problems, and there was an early belief among policy-makers (at least in Norway) that an international treaty limiting CO_2 emissions would soon be forthcoming (SSB 1989: 11). Such an agreement was clearly central to ensure that national emission reductions in Norway would *de facto* contribute to global emission reductions, rather than just be an extra burden on Norwegian business. The assumption among policy-makers that a treaty would be achieved helps explain why the Parliament adopted the stabilization by 2000 target.

The stabilization target could be reached through one of two methods: direct regulation, where industrial actors are given licences, technological standards and emission quotas, or a price incentive, e.g., taxation. The former method gives a high degree of target efficiency, whereas the latter is more cost-efficient. Putting a tax on pollution is an economics textbook solution, because it gives economic actors a way of "internalizing" the "externalities" of their activity whilst being a cost-efficient way of distributing pollution costs (putting a "price on carbon"). The cost of pollution is environmental degradation, and taxation is a way of ensuring that the "polluter pays".[4] The "polluter pays" principle has been important since the adoption of the UNF-CCC in 1992, and is also central in Norwegian environmental policy.

The choice of instruments in pollution control was largely a battle between the Ministry of Environment and the Ministry of Finance. The former preferred direct regulation (quotas, licensing) because they increased target effectiveness, thus increasing control with the amount of pollution, whereas the latter preferred economic incentives (taxes and subsidies) to improve cost-efficiency (Bretteville and Søfting 1998). The recent history of high unemployment in the late 1980s made politicians sensitive to situations that could weaken industry's international competitiveness. A unilateral Norwegian carbon tax would, it was argued, harm Norwegian competitiveness and consequently Norwegian jobs. Continued economic growth was seen as a necessary prerequisite for a strong welfare state. At the same time,

---

3   The Montreal Protocol from 1985 (CFCs), the Helsinki Protocol from 1985 and the Oslo Protocol from 1994 (SOx), the Sofia Protocol from 1988 (NOx), among others.

4   A tax on pollution might also act as an incentive to technological development. By pricing pollution, any technology that can be developed and implemented cheaper than the tax will be profitable. This principle is the basis of the numerous cries for "putting a price on carbon", so that industry would have a baseline against which investment in emission reduction technologies could be measured.

the Brundtland Commission's conclusions, that the global community had sufficient information to act to reduce the threat of climate change, weighed heavily on Norwegian politicians.

The Ministry of Industry had initiated a study in 1988, undertaken by Statistics Norway, resulting in the SIMEN report (SSB 1989), on how to follow up the Brundtland Commission's report "Our Common Future". The SIMEN study's main aim was to investigate whether it was possible to "pursue an ambitious environment policy and simultaneously achieve an acceptable economic growth", and furthermore, which policy instruments would be best suited to "modify possible target conflicts" (SSB 1989: 1). The report, the only macro-economic study of its kind at the time, showed that stabilization of CO_2 emissions was not incompatible with continued economic growth. Among the scenarios presented, the only one that would lead to a stabilization of CO_2 emissions at the required level was the one where taxes on fossil fuels were increased significantly (SSB 1989: 93).

In 1991, the Parliament adopted CO_2 taxes, as recommended by the available economic literature at the time. Initially the tax was applied to mineral oil and gasoline, but was later in the same year extended to include emissions from the petroleum production on the continental shelf. Whereas the petroleum industry thereby became subject to a tax of NOK 300 per tonne of CO_2 emissions, the energy intensive industry was exempt on the grounds of damages to its international competitiveness (and possible relocation abroad with an accompanying loss of jobs in Norway) (Gullberg and Skodvin 2011). The SIMEN report, providing the underlying rationale for the carbon tax, had made two crucial assumptions for its success, both of which failed to materialize: firstly, that economic growth in Norway happened mainly in the non-petroleum sectors of the economy, and secondly, that the carbon tax would be applied universally (Hovden and Lindseth 2002:148).

Throughout the 1990s it became increasingly clear that the cost of adhering to the 1989 stabilization target by 2000 was going to grow significantly both in terms of finances and employment. The late 1980s and the early 1990s was a period of liberalization in many sectors, in Norway spurred among other things by its negotiations for a possible membership in the EC. The previous goal of extracting Norwegian petroleum resources in a "moderate tempo" were effectively shelved with the liberalization in the petroleum sector. Emissions grew with increasing activity. According to a report from the oil association OLF in 1993, technology to reduce emissions from oil explorations was exceedingly costly. This in effect laid to rest the optimism about technological solutions for the emissions problem, an optimism that had been crucial for the legitimacy of the stabilization target (Hovden and

Lindseth 2002:148). The perceived choice was, effectively, between emission reductions *or* jobs.

Norwegian policy-makers deserve credit for trying to create better pre-conditions for their national policies. To work for a global system which secured that national cuts resulted in real emission cuts and not just relocation of economic activity was critical to the idea of an emission trading system. In addition, such a system could be a solution to the perceived choice between jobs and emission reductions. This would ensure the principle of cost-efficiency, while allowing Norway to continue its oil exploration as well as retaining its position as an environmentally responsible country.

A White Paper from the Ministry of Environment from 1995 lists economic cost-efficiency as the pivotal element in the choice of policy instruments in Norwegian climate policy. "An effective, international climate strategy should seek to identify cost-efficient solutions across countries, sectors and gases" (EMK 1996).[5] The logical consequence of this line of thinking was to increase efforts to reach an international agreement. Such a strategy became even more appealing by the fact that it offered a justification for continued Norwegian oil and gas export, not least because Norwegian oil was being promoted as "relatively clean" (when emissions from production were compared to other petroleum regions), and because gas would help reduce emissions if its use replaced the use of power from even dirtier coal-fired power stations.

As the negotiations on the Kyoto treaty progressed, the Norwegian position shifted from favouring a regime with national emission targets, to one where the flexible mechanisms and quotas trading were the main policy instruments. After the adoption of the Kyoto treaty in 1997, this was regularly portrayed as the main pillar of Norway's climate policy. Because the country was small and an open economy, so the argument went, an internationally binding agreement was a necessary precondition for an effective national climate policy.

Thus, by the end of the 1990s, it became clear that international emission trading had become preferred policy over direct regulation and domestic carbon taxes.

---

5   https://www.stortinget.no/no/Saker-og-publikasjoner/Publikasjoner/Innstillinger/Stortinget/1995-1996/inns-199596-114/?lvl=0#a1 (accessed 28 November 2014)

9.2 HOME OR ABROAD – WHERE TO REDUCE EMISSIONS?

GHG emissions are influencing the global commons that is our atmosphere. For atmospheric GHG concentration, it is irrelevant where emissions take place. Emission reductions' impact on national economies, however, as well as on national political situations, varies greatly across countries.

Norway's situation as a small and open economy with almost all inland electricity consumption covered by hydropower, a high degree of industrialization and technological development, makes inland emission reductions relatively costly compared to possible cuts elsewhere.[6] If cost-efficiency is the overall goal, one should target the cheapest emission cuts first, i.e. abroad. Furthermore, paying for emission cuts in other countries could also, and crucially, enable technology transfer to poorer countries, and contribute to poorer countries' economic development.

Arguments for national emission reductions are mainly political and moral: one cannot retain legitimacy as a climate policy pioneer while continuing to have high per-capita emissions. Norway is a rich country, and using some of this wealth on demonstrating that a low-emission society is not necessarily miserable, but rather a modern, efficient and technologically advanced society, would strengthen the incentive for others to follow suit.

This debate, which in many ways parallels the old debate between target effectiveness and cost-efficiency, has been important in Norwegian climate policy debates since the 1990s. The Kyoto treaty, adopted in 1997, provided an opportunity to bridge the divide between cuts "at home" and cuts "abroad".

Norway's obligations under the Kyoto treaty were to restrict average yearly emissions during the Kyoto period 2008–2012 to one per cent above the annual emissions in 1990. This implied that Norwegian emissions in the period 2008–2012 should not exceed 50.1 $MtCO_2e$/year. The Kyoto protocol stipulated that the countries that had taken on obligations to reduce their emissions, should do so "primarily through national measures",[7] and that the flexible mechanisms should be seen as a supplement. The flexible mechanisms were, however, to become much more important in Norway's fulfilment of its obligations than envisaged in the treaty.

By 2001, emission trading was easily the most important element in the government's climate policy, in addition to the carbon tax. Energy-intensive

---

6   An example is that in 2010 the Climate and Pollution Directorate estimated macroeconomic costs
      of measures to reduce emissions from private cars to be in the range of NOK 200 to NOK 4000
      per tonne CO_2e. As a comparison, Brazil's efforts to retain its tropical forests are being paid in the
      area of $5, or approximately NOK 30–NOK 40 per tonne CO_2e.
7   http://unfccc.int/kyoto_protocol/mechanisms/items/1673.php accessed 2 Dec. 2014.

industries had managed to be exempt from the carbon tax, and in return entered into "voluntary emission reduction agreements". Norwegian climate policy was based on pursuing an international agreement on emission reductions,[8] a continuation of the carbon tax, both for the petroleum sector and in the transport sector, voluntary agreements with industry, research and development, hereunder carbon capture and storage, and waste management to increase recycling and reduce landfill. The government also wanted to do "a reasonable share" of emissions reductions "at home" (as opposed through international emission trading), but conceded that it was "not useful to put a number on how much [of the emission reductions] that should be taken nationally (…) If we see significant technological progress in important areas, or new or alternative solutions to e.g., energy use are developed, this may lead to a very different national emission scenario."[9]

Whereas the centre-right government (2001–2005) signalled a wish to do a "significant share" of emission cuts at home,[10] their choice of policy instrument did not differ from the previous government's; they established a national quotas trading system already from 2005, rather than in 2008, which was the start of the Kyoto period.

As time progressed, the "home or abroad" question became increasingly difficult to answer. True, the advocates of national cuts regularly argued on moral grounds, as opposed to those arguing for cost-efficiency. However, as the Kyoto protocol entered into force, Norway joined the EU ETS, and 35–40 per cent of Norwegian GHG emissions were covered by the EU emission trading system. Consequently, any reductions that were done within the part of the economy subject to carbon trading would engender emission quotas that could be sold in the international market, and therefore not contribute to global emission reduction. As the EU ETS was extended by the end of the first period and Kyoto II entered into force in 2013, the proportion of Norwegian emissions covered by carbon trading increased to closer to 50 per cent. Such a development is positive if the carbon market is well-functioning. However, since the financial crisis in 2008, prices have remained too low to function as an incentive to invest in new, low-emission technologies.

At the same time, Norway's inclusion in the EU ETS has removed some of the national policy-makers' ability to regulate for emission cuts at home. By participating in the European carbon market, "home" is effectively enlarged to include the other 30 countries in the EEA.

---

8   The Kyoto treaty was not yet ratified, and was threatened by US withdrawal.
9   St.meld. nr. 54 (2000–2001): 21.
10   St.meld. nr. 15 (2001–2002): 2.

9.3   HOME AND ABROAD: THE GAS QUESTION

The issue of gas-fired power stations has been controversial in Norway since the 1980s, and led to the fall of the centre-right government in 2000. Gas reserves on the continental shelf are important, and relatively more so after the decline of Norwegian oil production since around 2000. Power stations are large and costly infrastructure, and have consequences for a country's ability to choose its energy provision over a long time period because of the huge sunk costs. Evaluations on a project's economic viability are therefore dependent on realistic scenarios for future energy needs and production capabilities.

In the 1990s projections showed that hydropower-based electricity production was approaching its practical limit, while electricity use continued to grow. Gas-based power plants could therefore alleviate a perceived looming electricity shortage, and to the extent that exported electricity would replace dirtier coal-based power abroad, it would also contribute to lower global emissions (Hoel and Strøm 2009). The main opposition to gas-fired power stations in Norway came from the environmental movement, and their base argument was the need to reduce, not increase, Norwegian GHG emissions (Quiviger and Herzog 2001).

Carbon sequestration technology was still in its early phases in the 1990s, expensive, and with no large-scale commercial operations. The National Pollution Authority (SFT) had in 1997 given two licences for gas-fired power stations (at Kårstø and Kollsnes), but the emission restrictions placed upon the licensees meant that the operation would be commercially unviable and necessitated technology that was not yet sufficiently developed. The issue became politically charged, and the minority government went to the Parliament for a vote of confidence, and lost in 2000.

The issue at stake was whether Norway should effectuate all (or nearly all) of its emission reductions "at home" (i.e. nationally), or whether increased national emissions were permissible provided they led to lower global emissions. The argument is closely linked to the existence of an international market for emissions trading. This is assumed to be a cost-efficient way of implementing emission reductions, as it theoretically ensures that the cheapest reductions are made first.

The incoming Labour government saw Norwegian gas-based electricity export as a bonus for the global fight against climate change, as it was assumed to replace coal-based Danish electricity production. In a deregulated international energy market, Norwegian power suppliers needed to be able to compete on the international arena. In order to protect Norwegian jobs, they should not be burdened with heavier regulations than their European counterparts.

Moreover, allowing gas-fired power stations with carbon sequestration was seen to contribute to the development of a technology that could be an important part of the global solution to the climate challenge. As the state became heavily involved in the financing of the new technology, effectively promising to carry the costs, politicians could claim to do something to reduce GHG emissions "at home" that would undoubtedly be beneficial also to the wider global community, while at the same time safeguarding jobs and Norwegian competitiveness.

Unfortunately, this particular political "triple-win solution" has so far proven not to deliver on its promise. Two of the three gas-fired power stations that have been built in Norway (at Kårstø, Mongstad and Melkøya) have so far been both economic and political failures. Higher than expected prices for gas and low electricity prices have resulted in the power stations being unable to sell electricity in the market at profit. The power station at Kårstø went online in November 2007, but has been mostly shut since then and completely closed down in October 2014.[11]

The gas fired power station at Mongstad was given a licence to operate in 2006, under the condition that CCS technology was being developed and applied as soon as possible, and no later than 2014.[12] The station was designed to deliver both heat and electricity to the next-door refinery, and any excess electricity would be delivered to the grid. Therefore, the Mongstad project was from the outset more about technology development and efficient production of electricity and heat in tandem, than had been the case at Kårstø.

Financing the technological development was largely to be carried by the state, through TCM (Technology Centre Mongstad). Statoil's financial burden in the establishment of the full-scale installation was limited to "the company's alternative carbon cost if CCS had not been implemented (carbon costs equal to other national industries exposed to international competition)".[13] This would mean, in practice, that Statoil would pay the price of the carbon quotas, regardless of whether full-scale CCS was installed or not, and the state would cover the risk. The Prime Minister Jens Stoltenberg called the Mong-

---

11   http://e24.no/energi/driften-av-gasskraftverket-paa-kaarstoe-opphoerer/23308662

12   http://www.regjeringen.no/nb/dep/kmd/tema/plan--og-bygningsloven/plan/kommuneplan-legging/innsigelsessaker/utvalgte-brev/2006/utslippstillatelse-for-co2-for-statoils-.html?regj_oss=1&id=448114

13   http://www.regjeringen.no/upload/kilde/oed/prm/2006/0143/ddd/pdfv/296763-internet-tomtale_-energiverket_pa_mongstad_og_avtale_mello.pdf

stad project "our moon landing" in his New Year's speech in 2007,[14] thereby investing much political capital and prestige in the project.

It turned out to become a costly affair. The scaling of cleaning technologies and safety issues regarding the actual building of the installation colluded with financial problems. The planned full-scale operation was postponed several times, and was finally shelved in September 2013, because of a "high risk level".[15] The technology centre, however, as opposed to the full-scale operation, has undoubtedly produced high-value knowledge and is co-operating with CCS projects elsewhere.

Full-scale CCS at Mongstad was one project where it can be argued that clean cost-efficiency measures were not pivotal to its initiation. Rather, the project has since the start been characterized by high political ambitions and prestige. The fact that this project failed might be seen as a timely reminder of the weight of the cost-efficiency argument in climate policies. There is an observable consensus that the organizational handling of full-scale CCS at Mongstad has been catastrophic, and the political process surrounding it has been full of intrigues and political positioning. However, it is an undisputed fact that TCM has brought important contributions to the global development of CCS technology. The passing of time may pay more heed to this aspect, and less to the surrounding political wrangling.

9.4 EMISSION REDUCTION POTENTIAL AT HOME
   – OVERARCHING POLICY TARGETS
   AND INSTRUMENTS IN SOME SECTORS

So far this chapter has looked at two "meta-issues", namely the role of cost-efficiency and the question of home vs. abroad emission reductions. The remainder of the chapter will discuss the development of national targets for emission reductions as they have appeared over time.

Whereas the theoretical and to some extent also practical work on a global trading system was advanced and refined throughout the 1990s, inland emissions remained stable and above the 1990 level. As shown above, Norway's position as a small country with an open economy had been used as an argument that national emission reductions would be insignificant in the global

---

14   http://www.regjeringen.no/nb/dokumentarkiv/stoltenberg-ii/smk/taler-og-artikler/2007/
      statsministerens-nyttarstale-2007.html?id=440349

15   http://www.regjeringen.no/nb/dokumentarkiv/stoltenberg-ii/md/Nyheter-og-pressemelding-
      er/nyheter/2013/legger-om-arbeidet-med-fangst-og-lagring.html?regj_oss=1&id=735972

context. However, the obligation to reduce emissions per capita is seen as both moral and important for international political legitimacy.

But even if most national cuts are not cost-efficient in an international context, they may be more or less so within the national setting. How to determine whether to spend money and resources on CCS or low-emission vehicles? Would it be better to legislate for more recycling or better insulated buildings? Should we leave the Norwegian forest standing as carbon storage, or should we use it to heat our homes and fuel our cars?

To answer such questions, bureaucrats and experts from academia and industry started compiling and analysing information. Since the turn of the century, a long list of reports and documents has been produced to show the nominal and relative cost and emission reduction potential of various policy measures. These reports are meant to clarify the situation so that "rational" and cost-efficient policy decisions can be taken. Unfortunately, even if they repeatedly have shown the existence of a non-negligible emission reduction potential with macro-economic costs below zero, and an important emission reduction potential with relatively low macro-economic costs, the Norwegian emission curve has not bent downwards yet.[16]

The National Pollution Authority (SFT) published its first "effect analysis"[17] in 2000, and later updated it in 2005 and 2007. These documents gave a cost estimate for a large number of possible emission reduction measures, again adhering to the principle of cost-efficiency in a national setting.

In 2005, the Low Emission Commission (*Lavutslippsutvalget*), led by Jørgen Randers, was given the task of examining how Norway could reduce GHG emissions by 50–85 per cent relative to 1990 by 2050 (Lavutslippsutvalget 2006). The commission concluded that the task was feasible, inexpensive, and urgent. They recommended 15 policies, including public information campaigns, technology development, new renewable energy, transport both on land and at sea, energy efficiency in buildings, in the electricity infrastructure and in industry, and CCS. The Commission estimated the cost of these necessary measures to be "negligible" in a macro-economic perspective towards 2050 (Lavutslippsutvalget 2006: 105).

The Commission was not asked to evaluate or recommend policy measures, but did nevertheless make three recommendations. Firstly, that sector

---

16 One should, in the interest of balance, mention that the policy instruments introduced since 1991 have contributed to lower emissions than what would otherwise have been the case. Furthermore, some of the policies introduced since 2007, such as new building regulations, will not have any observable effect on emission reductions for some time.

17 "tiltaksanalyse".

wide action plans be developed, secondly, that the system for carbon trading be further developed and extended to include more sectors and countries, and thirdly, that public procurement be used more effectively to encourage low-carbon solutions.

The subsequent White Paper "Norwegian Climate Policy"[18] from 2007 outlined three goals for the government's climate policy:

- Norway shall become "carbon neutral" in 2050.
- Norway shall until 2020 be committed to cutting global emissions with an amount equal to 30 per cent of Norwegian emissions in 1990.
- Norway shall exceed its own Kyoto obligations by ten percentage points, to nine per cent lower than 1990 levels rather than one per cent above.

Common for these three goals, of course, is that they can be fulfilled through buying emission quotas in the international carbon market, and that they say nothing about national emissions.

It is worth remembering that at this time climate policy was high on the public agenda. In the UK, Sir Nicholas Stern had published the influential report "The Stern Review Report on the Economics of Climate Change". The IPCC had published its fourth Assessment Report, and together with Al Gore they were awarded the Nobel Peace Prize in 2007. Hopes for a global, binding emission reduction agreement in Copenhagen in 2009 were mounting. The media attention to climate policy was higher in 2007 than in any of the preceding years.

The White Paper triggered the parliamentary inter-party Climate Agreement (*Klimaforliket*[19]) between six of the seven political parties represented in Parliament in January 2008.[20] This was in itself an important achievement, and created a more stable foundation for national climate policy. A cross-party agreement ensured a continuation of the agreed goals and targets, regardless of which party holds governmental responsibility.[21]

In addition to demonstrating cross-party support, the Climate Agreement confirmed the fundamental principles of "polluter pays" and of cost-effi-

---

18   St.meld. nr. 34 (2006–2007) Norsk Klimapolitikk.

19   http://www.regjeringen.no/Upload/MD/Vedlegg/Klima/avtale_klimameldingen.pdf

20   The Progress Party were not invited to negotiate on or sign the Climate Agreement, which was mostly because the party until 2007 and later was the only political party sceptical about the IPCC conclusions, but partly also an element in the political game of power and influence.

21   This is clear to see in the post-2013 coalition government, where the Progress Party has signed onto a political platform that on several occasions refers to the parliamentary Climate Agreement.

ciency. It also settled the decades-old question of "home or abroad" by stipulating that two-thirds of the projected cuts until 2020 were to be achieved nationally. The final agreement had tightened the government's proposal from 13–16 MtCO2e below the reference path[22] to 15–17 MtCO2e below the path, giving a political target of emissions in 2020 not exceeding 43–45 MtCO2e. In addition, the Climate Agreement promised increased focus on

▶ new technology, both for CCS and for new, renewable technologies
▶ emissions reductions from the transport sector

Combining such political ambitions with overall cost-efficiency in a situation where a large and growing part of Norwegian emissions were covered by the EU ETS was a challenging task. The bureaucracy was put to work to try and identify the relative cost of different climate policy measures. The largest and most exhaustive of such exercises to date was published in February 2010, under the title of "Klimakur 2020" (*Climate Health* 2020). The report synthesized results from a broad range of studies in many sectors, analysing over 160 different measures. Each of these measures had an estimated cost and emission reduction potential. The report also discussed opportunities and barriers to implementation.

The report never became the "Road Map" hoped for by many. It did, however, succeed in demonstrating the large variation both in prices and emission reduction potential, and would become one in a long and growing line of reports about climate policy in Norway.

Yet another White Paper on Climate Policy was published in 2012. In addition to citing sustainable development and the "polluter pays" principle as the two guiding principles of Norwegian climate policy, the White Paper restated the overarching goals from 2006 about exceeding the Kyoto obligations, contributing to cutting global emissions until 2020 with an amount similar to 30 per cent of Norwegian 1990 emissions (increasing to 40 per cent if this could contribute to reaching a binding global agreement), and being carbon neutral in 2050.

Two elements were new this time around. Firstly, that the government stated that its long-term goal was for each global citizen to have an equal right to emit greenhouse gases. This element originated in the political platform hammered out between the three parties[23] before they formed a coalition government in 2009. Secondly, the general pessimism surrounding the inter-

---

22   As presented in the National budget for 2007.
23   The Social Democratic Labour Party, the Centre Party, and the Socialist Left Party.

national carbon trading arrangements, low carbon prices, and the slow progress in CCS, led the government to propose stronger measures to cut national emissions. The main policy suggestions were:

1. Increase the CO_2 tax in the petroleum sector (in effect, restoring it closer to the level it had been before the sector joined the EU ETS).
2. Increase the potential for use of electricity rather than gas-fired turbines on offshore installations.
3. Establishing a "climate and energy" fund to finance research and development into new low-carbon technologies and energy efficiency.
4. Tighten building regulations to lower energy use in buildings.
5. Reduce emissions from the transport sector through better public transport, encourage use of cycling and walking, increased use of biofuels (subject to sufficient sustainability criteria), and encourage electric vehicles through taxation and subsidies.

The Climate Agreement from 2008 had not only secured a consensus between six of the seven parliamentary parties, it had also cemented the divide between the Progress Party and the rest. The Progress Party has on several occasions expressed annoyance that they were "not invited" to the climate agreement negotiations in 2007. This cleavage became visible in the debate on the White Paper on Climate from 2012, where the Progress Party's representatives put on record that they saw climate change as one of several important challenges for humanity, and that scaremongering from politicians and the environmental movement was unhelpful. They were "sceptical about a debate that categorically assumes that changes in temperature and climate were human-made, and that solely bases environmental measures on reduction of CO_2 emissions rather than discussing necessary adaptation to the natural climate variation that nevertheless takes place".[24]

The 2012 White Paper led to an updated Climate Agreement between six of the seven parliamentary parties, again without the Progress Party's participation. The observable trend in the decade since the turn of the century had been a rhetorical shift back from international carbon trading and global cost-efficiency towards a stronger re-emphasis on possibilities for national reductions. This trend was followed in the new Climate Agreement from 2012, which reiterated the goal of making two thirds of the country's emission reductions nationally. The 2012 Climate Agreement further strengthened

---

24 https://www.stortinget.no/no/Saker-og-publikasjoner/Publikasjoner/Innstillinger/
Stortinget/2011-2012/inns-201112-390/2/

the government's proposal by increasing the capital in the new climate and energy fund, now renamed the "climate, renewable energy and energy transition fund", tightening regulations regarding energy use in buildings, and continuing to strengthen public transport, and to encourage use of low-emission vehicles through direct regulation as well as economic incentives in the transport sector.[25]

9.5   RECENT DEVELOPMENTS:
THE CARBON BUDGET AND DIVESTMENT

"Carbon budget" is a term that was originally used in natural sciences to describe the totality of accumulation into the earth system's carbon cycles (see Sterman this volume). In recent times, it has been adopted by social scientists and policy-makers[26] to mean the maximum amount of greenhouse gases that can be released into the atmosphere if the chance of preventing dangerous climate change shall be limited to a certain level.[27]

Two reports from 2012/2013 put the term on the international agenda, with a marked impact on Norwegian climate policy:

▶ The IEA's Energy Outlook 2012, stating that "[n]o more than one-third of proven reserves of fossil fuels can be consumed prior to 2050 if the world is to achieve the 2°C goal, unless carbon capture and storage (CCS) technology is widely deployed" (IEA 2012: 3).

▶ HSBC's "Oil and Carbon Revisited", looking into the value at risk from "unburnable reserves", concluded among other things that Statoil was the European company most overvalued, in the amount of 17 per cent of its market capitalization (HSBC 2013: 2).

The reports not surprisingly made large headlines, but even more importantly, they spearheaded a new approach to the climate challenge, in a language that would potentially engage broader parts of society.

Some large institutional investors, such as pension funds, have sold out of coal and tar sand companies, and in 2013 the Nordic countries and the USA

---

25   https://www.stortinget.no/no/Saker-og-publikasjoner/Vedtak/Vedtak/Sak/?p=52754

26   Carbon Tracker has done pioneering work in this area. See www.carbontracker.com.

27   The current political ambition is not to increase global temperature by more than two degrees Celsius compared to pre-industrial times (the 'two-degree target'). The probability of meeting this target varies with varying levels of GHG concentration in the atmosphere.

announced their plans to stop investing in new coal companies overseas.[28] The divestment trend seems to be spreading, even if its financial impact is still limited.

As of the parliamentary elections in 2013, Norway has had a coalition government of the Conservative Party and the Progress Party. The co-operation agreement between the two parties states that "the climate challenge is global and can best be solved globally".[29] The agreement clearly shows that international carbon trading still is the government's preferred instrument, and that the petroleum sector is envisaged as a crucial part of the Norwegian economy for "the foreseeable future". The minority government has promised a White Paper on energy, they will contribute to a cost-efficient CCS technology, and they will continue the subsidies for low-emission vehicles until 2017 and for biofuels until 2020.

At the time of this writing the right-leaning government's climate policy is a rather vague concept. The Conservative Party has signed the parliamentary Climate Agreement, whereas the Progress Party has been seen as a "climate policy pariah" whose voters are unconvinced by scientific facts about climate change. Much emphasis has been placed on engaging industry and society at large in the switch to the "green economy", with a series of public meetings between high-level politicians, industry, scientists, NGOs and the general public. However, the government has been heavily criticized for not putting their money where their mouth is. Their mantras of less regulation, less taxation, less bureaucracy, more individual freedom and personal choice are not easily reconcilable with a target-effective climate policy that will have a significant effect on the Norwegian emissions level.

But while the government shows signs of wanting to continue a high activity level in the petroleum industry,[30] Parliament has become more active on the climate policy arena. The recently defeated Labour party suggested in autumn 2013 a public report on the Petroleum Fund's investments in coal companies, the Green Party suggested divesting from coal and tar sands, and the Socialist Left Party suggested phasing out investments in fossil energy.[31] Clearly, it seems easier to lead an aggressive climate policy from the opposition benches.

---

28   http://www.washingtonpost.com/blogs/wonkblog/wp/2013/06/27/the-u-s-will-stop-subsidizing-coal-plants-overseas-is-the-world-bank-next/

29   "Politisk platform for en regjering utgått av Høyre og Fremskrittspartiet", p. 61.

30   See, for example, the Petroleum and Energy Minister's speech to the annual conference in Norwegian Oil and Gas http://www.regjeringen.no/nb/dep/oed/aktuelt/taler_artikler/minister/taler-og-artikler-av-olje--og-energimini/Norsk-olje-og-gass-arskonferanse.html?regj_oss=1&id=753394

31   https://www.stortinget.no/no/Saker-og-publikasjoner/Publikasjoner/Representantforslag/2014-2015/dok8-201415-042/

CONCLUSION

The history of climate policy in Norway from 1990 until 2014 is one of hot debates and issues, but continued high GHG emissions. This chapter concludes with three observations.

Firstly, some fundamental principles have remained throughout the period under investigation. These are the "polluter pays" principle, and the quest for cost-efficiency. They were central in the early 1990s when the CO_2 tax was first introduced, and they were central in the last Parliamentary Climate Agreement. Whether different politicians at different points in time interpret these principles in a similar fashion is a relatively open question. These concepts do, however, allow observers and participants of all kinds to make sense of new decisions in the context of decisions past, and they help to construct a continuous narrative through which Norwegian climate policy seems to be built on a stable base.

Secondly, more than two decades after the Norwegian parliament adopted the stabilization target for GHG emissions, we have greatly increased our knowledge about potential for emission reductions from inland sources. We have built statistics and indicators around a segmentation of human activity that contributes to framing the debate about climate policy. Groups of economic activity are being produced and reproduced, hinting at societal connections that may or may not be the most helpful to initiate the behavioural change that is needed to achieve a low-carbon society. Unfortunately, this knowledge has not yet resulted in significant emission reductions.

The third observation is that Norwegian climate policy is continually drawn between lofty intentions and *realpolitik*, between the need for jobs and the side effects of an active fossil energy sector. In the case of global warming, the Norwegian petroleum industry must definitely be considered among the polluters. However, the sector's importance for Norwegian wealth and job creation has made it all but impossible for politicians who want re-election to impose policies that would seriously threaten its existence.

Recent attempts to reframe the issue, by changing the time horizon for public investments, by highlighting the economic as well as environmental risks in the fossil energy sector, and by debating legitimate investments for the Petroleum Fund, have contributed to a slight change in public debate. Several pension funds have also advocated de-investing in companies producing energy from coal, thus contributing to the "destruction of demand" voiced as a possible way to end the fossil era. The threat (and reality) of a significant number of job losses in fossil energy-related sectors is likely to be much more effective in developing a low-carbon Norway than any number of cost-efficiency analyses. The dramatic fall in oil prices since the summer of 2014 has also shown how dependent the Norwegian economy is on fossil fuel exports.

It is in many ways delightfully ironic when the Finance Minister, who is also the Progress Party's leader, explains that what we see is the beginning of the necessary *and anticipated* restructuring of the Norwegian economy away from fossil fuel dependence.

In this way, the best hope for the future may be that the *Realpolitik* forces the politicians' hands to ensure real sustainable development. Unfortunately, the hurdles are still many.

BIBLIOGRAPHY

Bretteville, C. and Søfting, G.B. (1998). Beretningen om et varslet avgiftskutt. *Cicerone*, 7(7). http://www.cicero.uio.no/cicerone/98/7/cicerone9807.pdf

EEA (2014). *Annual European Union greenhouse gas inventory 1990–2012 and inventory report 2014*. Technical Report no. 9.

EMK (1996). *Innstilling fra energi- og miljøkomiteen om norsk politikk mot klimaendringer og utslipp av nitrogenkoksider (NOx)*. https://www.stortinget.no/no/Saker-og-publikasjoner/Publikasjoner/Innstillinger/Stortinget/1995-1996/inns-199596-114/?lvl=0#a1

Gullberg, A.T. and Shodvin, T. (2011). Cost effectiveness and target group influence in Norwegian climate policy. *Scandinavian Political Studies*, *34*(2), 123–142.

Hoel, M. and Strøm, S. (2009). Klimapolitikk for en liten, åpen og rik økonomi, in *Nytt Norsk Tidsskrift, 3-4*, 496–502.

Hovden, E. and Lindseth, G. (2002). Norwegian Climate Policy 1989–2002, in W.M. Lafferty, M. Nordskag and H.A. Aakre (Eds.), *Realizing Rio in Norway: Evaluative Studies of Sustainable Development*. ProSus report, University of Oslo.

HSBC (2013). Oil and Carbon Revisited. Accessible at: http://www.longfinance.net/images/reports/pdf/hsbc_oilcarbon_2013.pdf

IEA (2012). *Energy Outlook*. http://www.iea.org/publications/freepublications/publication/English.pdf

Lavutslippsutvalget (2006). *Et klimavennlig Norge*. NOU 2006: 18.

Quiviger, G. and Herzog, H. (2001). *A Case Study from Norway on Gas-Fired Power Plants, Carbon Sequestration, and Politics*. http://sequestration.mit.edu/pdf/netl_quiviger.pdf

Sprinz, D. and Vaahtoranta, T. (1994). The Interest-Based Explanation of International Environmental Policy, in *International Organization*, *48*(1), 77–105.

SSB (1989). *SIMEN: Studies of Industry, Environment and Energy towards 2000*. Discussion paper no 44.

CHAPTER 10

Norway's Role in Climate Negotiations since Kyoto

By Hanne Inger Bjurstrøm[1]

1   I would like to thank former chief climate negotiators, Harald Dovland and Audun Rosland, for sharing with me their thoughts and experience from their committed involvement in the negotiations under UNCCCC

10.1 NORWAY, A STRONG ACTOR IN THE NEGOTIATIONS, – BEFORE KYOTO AND TODAY

Since the adoption of the United Nations Framework Convention on Climate Change (UNFCCC) in 1992, Norway has been very actively involved in the work under the convention. Norway has dedicated experts and given broad financial support to the different ongoing processes. At the annual COPs (Conference of the parties under the UNFCCC and later also the Kyoto protocol MOPs), as a rule, Norway has been represented at ministerial level, in contrast to many other countries. Hence, Norway has been given a fairly influential part in the negotiation process, relative to our size and share of the total global emissions. In practical terms, this is reflected in the many working groups and subgroups that have been, and still are, co-chaired by delegates from Norway. This position cannot be fully explained by the fact that Norway has provided money and qualified people only. It must also be seen in the context of Norway's overall engagement in foreign policies.

In 1983 Gro Harlem Brundtland was appointed as chair of the World Commission on Environment and Development. The work resulted in the "Brundtland report," a cornerstone in the development of the concept of "sustainable development." Under the notion of "our common future," the report broadly advocates the need for united global efforts in pursuing a sustainable future. The report emphasized the strong link between the environmental perspective and economic growth. As early as in 1972, another Norwegian, Jorgen Randers, was co-editor of the report "The Limits to Growth." Both of these publications have to a great extent influenced how the global society has viewed the issue of climate change.

Furthermore, Norway was a frontrunner internationally in adopting policy instruments to reduce CO_2 emissions nationally. Norway was one of the first countries in the world to adopt a CO_2 tax (1991), and to implement green tax reforms.

In addition to this, Norway's financial contributions to the UN, and other international bodies and processes are well known internationally. Norway's record as a substantial provider of funding for developing assistance in general, and to countries which are very vulnerable to climate change in particular, also enhances dialog and trust. Norway is to some extent seen as a country that does not act according its own interests in the negotiations, although the veracity of this perception is debatable.

Not being an EU-member also leaves Norway in the unique position of being a country with an independent voice in the negotiations. Not having to coordinate positions with a group of countries makes it easier to develop new compromises and proposals on the spot when the negotiations are stuck.

This might explain why Norwegian negotiators are often chosen to chair meetings under UNFCCC. Further, this situation also explains the fact that Norway is involved in many other processes linked to the negotiations, but formally outside the UNFCCC. Norway has frequently been invited as an observer to the MEF meetings (Major Economies Forum, initially the G20 countries) and a host with the EU, on a regular basis, of meetings with large groups of countries.

NORWAY'S POSITIONS IN THE NEGOTIATIONS UP TO KYOTO

The Norwegian positions in the negotiations up to Kyoto were founded on some specific elements and conditions featuring Norway. These are significant, as they are still viable, and thus still guide Norwegian policies in the negotiations – also post–Kyoto.

Firstly, since Norway is a small country with a relatively small share of the total global greenhouse emissions, a global approach to the problem is clearly in Norway's interests. Further, Norway has a tradition of a strong commitment to the UN and its principle of "one country one vote." Thus, coordinated actions under the auspices of the UN are in line with our general foreign policy.

Secondly, the Norwegian position is a science based approach to the climate issue. Hence, we align our climate policy with the findings of the IPCC (Intergovernmental Panel on Climate Change). The Norwegian policy has been linked to the two degree goal since it was introduced by the IPCC (the goal limiting the increase in the global average temperature to 2 degrees Celsius above the preindustrial level). The level of ambitions in a global agreement has to respond to what is needed for keeping the world on an emission pathway in line with this goal.

Thirdly, Norway is an oil and gas producing country. Approximately one third of our national emissions of greenhouse gases derive from the oil and gas sector. The CO_2 tax has proven to be an effective tool for reducing the emissions from this industry. According to the latest report from The Norwegian Environment Agency (Miljødirektoratet), the emissions from this sector would have been significantly higher without a tax on CO_2. Still, a substantial part of Norway's emissions derives from the production of oil and gas, in addition to large revenues and many jobs. On the other hand, almost 100% of our power consumption domestically (offshore excluded) is based on hydro power. Norway cannot, unlike the EU, reduce our emissions by changing the mix of our energy supply, such as by shutting down a coal power plant.

This is one of the reasons why Norway has been a strong supporter of the "off-set mechanism" during the Kyoto Protocol negotiations. Such a mechanism derives from the fact that climate is a global problem, and as long as emissions are reduced, it doesn't matter – in physical terms – where the reductions take place. The main principle is to have a system which is cost effective; hence emissions should be first cut where the marginal cost for cutting them is at the lowest.

These three features mentioned above are still guiding Norway's policies and positions in the climate negotiations.

THE KYOTO PROTOCOL AND NORWAY

The final Kyoto protocol and the subsequent decisions under the protocol did, to a large extent, meet the requirements put forward by Norway in the negotiations. The lack of success of the Kyoto protocol is due to the fact that the US withdrew its participation as a party to the protocol before it went into force. Further, many of the countries with fast growing emissions didn't have reduction targets under the protocol. Hence, the protocol, from the very outset, could not deal with the global scale of the problem at hand.

Leaving the geographic shortcomings aside, the basic elements of the Kyoto protocol were fully in line with Norwegian positions. The protocol established an overall cap on emissions (though not on a sufficiently large share of the growing global emissions) and "offset" mechanisms were established. Further, Norway had been a strong advocate for differentiated commitments for each country, opposite to a flat emission reduction rate allocated to all committed parties. The adoption of individually set targets, related to a common base year, thus fully accommodated Norway's positions. A third point which had been emphasized by Norway in the negotiations was the inclusion of forests (only "national forests," thus excluding forests in developing countries) in the scope of the Kyoto protocol. The protocol, with subsequent decisions, establishes a system for the inclusion of national forests in the emission accounts of a Party.

One should note that the respective positions of Norway and the EU differed on all three of these aspects of the agreement. Norway was at that time (in the 1990s) more aligned with the US. While Norway had a tradition of using fiscal policies for curbing emissions, in the EU (for obvious reasons the EU did not have a common fiscal policy at that time), regulatory measures dominated. Hence, the use of market mechanisms and offset rules were something the EU was reluctant about from the beginning. Further, the EU favored a flat rate commitment type allocated to all Parties and opposed the inclusion of national forests in the agreement.

It is the current and projected emission trajectories that explain the different approaches to the negotiations of Norway and the EU at that time. With

1990 as a "base year" for emission reduction targets in the Kyoto protocol, the EU (as a "bubble") would get a huge portion of its reduction obligations for "free." This was due to industrial reconstructions in the new member countries, which led to a decline in greenhouse gas emissions. However, Norway was at that time familiar with the domestic marginal cost of emission reductions, on the basis of studies performed by The Norwegian Pollution Control Authority (Statens Forurensningstilsyn). Hence, an offset mechanism, and the inclusion of national forests in the agreement, were regarded as important elements for reaching the emission reduction target set for Norway, both de facto, and according to cost effectiveness.

It is a paradox however, that in the end the US withdrew from the protocol whilst the EU, after having first opposed an emission trading system, developed the EU Emission Trading System (ETS). Norway is now, since 2000, in most parts closely aligned with the positions of the EU in the negotiations. Although a non EU country, Norway is still a member of the so-called Umbrella Group (UG), which was formed on the basis of the different country positions in the Kyoto negotiations.

The UG group comprises a very wide and different range of countries, such as the US, Russia, Australia, Canada, Japan, New Zealand, and Norway. NGOs frequently question how Norway, as a progressive country in the negotiations, can fit in with this group of countries. Apart from Australia, none of these countries are Kyoto members, thus strongly disconnected from Norway's positions in the negotiations. However, this group does not constitute a common policy platform, and less often forwards common statements within the negotiations. Thus, the group is a source of information and knowledge. It can be argued that Norway is in a unique position, in being able to exchange views and propositions with leading countries in the negotiations on a regular basis (during the sessions under UNFCCC, every morning).

10.2 THE UNFCCC AND NORWAY SINCE KYOTO 1998

THE KYOTO 2

At COP 13 (Conference of the Parties) in Bali in 2007, the "Roadmap to Copenhagen" was adopted. An extension of the Kyoto protocol into a new commitment period was a part of this road. In Durban in 2011, the parties agreed on the so-called Kyoto 2.

For the future negotiation process, it was important that the Parties managed to land an agreement on a Kyoto 2. If not, the negotiations would have been back to square one, and the agreements and mechanisms under the Kyoto protocol would have ceased to exist. For Norway, advocating strongly for similar types of mechanisms in a new agreement, this was important. Further, it would have been a very hard blow to the ongoing negotiation for a global agreement covering all countries, if a second commitment period under the Kyoto protocol had not been adopted. Yet, if one looks at the number of countries that have signed the Kyoto 2 – the EU, Norway, Switzerland and Australia, and the fact that countries like Canada, Russia and Japan are not parties to the protocol – it becomes clear that the protocol's contribution to trust building is rather weak. Further, for the climate, the protocol is of minor importance, currently covering 10–12% of the global emissions, and declining. Hence, the focus now within the negotiations is the so-called "Durban platform," a platform which paves the way for a global agreement covering all parties.

NORWAY IN THE NEGOTIATIONS IN THE
PERIOD BETWEEN BALI AND COPENHAGEN

Two initiatives, though not of equal importance, were launched by Norway in the negotiations before COP 15 in Copenhagen. The most important one was the launch of the "Norwegian International Climate and Forest Initiative" at COP 13 in Bali 2007. But also the so-called Norwegian proposal which was gradually developed in this period was the focus of much interest on the part of many countries and NGOs, thus positively influencing Norway's position in the negotiations.

THE "NORWEGIAN PROPOSAL"

The Norwegian proposal outlined an approach to raising funding for emission reduction measures in developing countries. It follows from the UNFCCC that the industrially developed ("Annex1") countries have an obligation to provide financial support to developing ("non-Annex1") countries for such measures. This issue has always been (and still is) heavily emphasized by the developing countries. They argue that sufficient money has never been put on the table by the developed world, and thus their part of the agreement has not been fulfilled. The Norwegian proposal was founded on a Kyoto-type global agreement, in which an overall global emission target for a certain period was set, with tradable emission allowances. A certain amount of the allowances should then be withheld from the distribution to the parties, and instead be auctioned globally. The income from these auctions should be allocated to a fund. Money from the fund would in turn be distributed, under certain rules, to activities for emission reductions in developing countries. The beauty of the proposal was the way it could raise money from "within the system itself." Hence, it would create funding, independent of the national budget processes within the member states. The weak part of the proposal, however, was that it depended on the adoption of a Kyoto-type agreement. Because of this, it never got real support from countries that had not signed the Kyoto protocol.

As time went by, the negotiations after 2000 increasingly moved in the direction of a so-called bottom-up type of agreement, thus making the Norwegian proposal less relevant. However, the proposal was partly picked up by Mexico in its proposal for a "Green fund," which was discussed at COP 15 in Copenhagen. Thus, there is a clear link between "the Norwegian proposal" and the fact that Norway, through Jens Stoltenberg, was given an important role by the UN Secretary General after Copenhagen, in the work of developing climate change financing.

NORWAY'S INTERNATIONAL CLIMATE AND FOREST INITIATIVE

This initiative was launched by Jens Stoltenberg in Bali. It aims to support efforts to slow, halt and eventually reduce greenhouse gas emissions from deforestation in developing countries (REDD Plus). Approximately NOK 3 billion is allocated to the project annually. The current government has announced that they will continue the project at least at the same level up to 2020. Combating deforestation and forest degradation is regarded as being one of the most significant and cost-effective ways of fighting climate change, to help keep the world on a 2-degree track. In addition, forest preservation in developing countries is a tool for reducing poverty and biodiversity decline.

The initiative has the following three objectives; 1) to contribute to the inclusion of reduction of greenhouse gas emissions from forests in developing countries under the UNFCCC, 2) to contribute to early actions for measurable emission reductions from deforestation and forest degradation and 3) to promote the conservation of primary forests, due to their particular importance as carbon stores and for their biological diversity.

The Climate and Forest Initiative has established a series of partnerships with key forest countries, such as Brazil, Indonesia and Guyana. Brazil, currently the predominant country of the three, can serve as an illustration of how the initiative basically works. In 2008, Brazil created the Amazon Fund to generate additional result-based financing to promote reduced deforestation in the Amazon. Norway has pledged up to one billion US dollars to the Fund by 2015, if Brazil demonstrates continued reductions in deforestation. Payments from Norway in a particular year will depend on the difference between emissions from deforestation in the previous year and a reference level. (The reference level is the average deforestation for a selected ten-year calculation period, updated every five years.) Brazil will only get payment according to documented results. Hence this is a "pay for performance" scheme. The Initiative also supports country programs for reducing deforestation through organizations such as UN-REDD and Forest Carbon Partnership Facility.

An important step for the inclusion of emissions from forests in developing countries under the UNFCCC was taken at COP 19 in 2013, with the adoption of "The Warsaw Framework for REDD Plus." The framework encompasses seven decisions related to REDD Plus, and is a significant step forward for creating a mechanism for rewarding developing countries for not clearing forests under the UNFCCC. Even if there are still many issues to be worked out, substantial progress was made on key issues like finance, transparency and safeguards, monitoring, verification, institutional arrangements, and how to address the drivers behind deforestation.

From the very onset, Norway has been a driving force in the work for obtaining progress on this part. The expertise and knowledge we have gained from the Climate and Forest Initiative, through our partnership with key forest countries, have given us a valued position in the negotiations. Thus, Norway has contributed significantly to the design and architecture of the model for REDD Plus, as seen in the adopted framework.

It is fair to say that the Climate and Forest Initiative has enhanced Norway's influence in the negotiations and not only on issues related to deforestation, but also in other areas. Channels for the exchange of interests and views have been created far beyond the partner countries. Among other things, outreach to other potential donor countries has created a platform for dialog between Norway and key countries in the negotiations. Further, the experiences with result-based funding gained by the Initiative have received broad support, and not only related to deforestation.

10.3 THE ROAD FROM COPENHAGEN TO COP 21 IN PARIS
FROM COPENHAGEN TO PARIS 2015

The poor outcome of COP 15 in Copenhagen was a shock to all those in the world engaged in climate policy. For years, the IPCCC, state leaders and NGOs had emphasized the urgent need for concerted actions through the adoption of a global, legally binding agreement in Copenhagen. For the delegates involved in the negotiation process, the failure of Copenhagen hardly came as a surprise. China and India had made it very clear that they were not ready to take such a commitment upon themselves. Here developed countries had to scale up both climate finance and emissions reductions, before they could ask anything from developing countries. For the US, (and Canada, Japan and Russia) this was unacceptable; all countries had to have obligations under a new agreement.

After Copenhagen, there was a need for a new platform for negotiating a global comprehensive agreement under the UNFCCC. This was adopted in Durban in 2011. Here, the Parties committed to "a new platform of negotiations under the Convention to deliver a new and universal greenhouse gas reduction protocol, legal instrument or other outcome with legal force by 2015 for the period beyond 2020, where all will play their part to the best of their ability and all will be able to reap the benefits of success together."

At COP 19 in Warsaw in 2013, a decision was adopted which invites all Parties (ready to do so) "to communicate their intended national determined contributions (INDCs) to the secretariat by the first quarter of 2015." The intention is to make sure that the parties start the process of estimating what their contribution to a Paris-agreement would be well in advance of December 2015. The INDC approach in itself illustrates that the possible result in Paris will be a so-called bottom-up type of agreement. While a top-down agreement would set a global target for the overall emissions, and allocate reduction targets to each individual country accordingly, the global emission reduction target in a bottom-up agreement would be the sum of the national reduction efforts. This differs from the approach of the Kyoto protocol, and the type of agreement the world tried to obtain in Copenhagen in 2009. Learning the lesson from Copenhagen, this seems to be a more realistic approach, but at the same time it mirrors the lower level of ambitions for the Paris COP.

NORWAY'S POSITIONS IN THE NEGOTIATIONS TOWARDS COP 21 IN PARIS

Norway's overall ambition for Paris is to obtain a science based and legally binding agreement that applies to all Parties. Norway is flexible on the type of document to be adopted, whether this is a protocol or a separate agreement, as long as it is legally binding on all Parties. However, at the moment obtaining a legally binding outcome in Paris seems to be very difficult, at least when it comes to legally binding emission reduction targets. The US for instance, is strongly against this, but seems to accept the adoption of legally binding rules on MRV (measuring, reporting and verification). Hence, the outcome in Paris would most likely be an agreement that is partly legally binding on the Parties (e.g., common rules on MRV, governance, structure etc), and partly based on voluntary national determined contributions (emission reduction targets).

A science based agreement means that the level of ambitions for mitigation efforts must reflect the findings of the IPCC. Hence, it must ensure that the world gets on track with respect to the 2 degree pathway. The gap between the current global emission trajectory and the reductions that are needed in 2030 and 2050 to ensure that the global average temperature doesn't exceed

2 degrees, has to be filled or at least substantially reduced. Further, the agreement must be flexible and dynamic, hence able to respond to new and additional findings by the IPCC. Norway has recently, as one of an increasing number of countries, argued for the inclusion of a long-term goal in the Paris agreement of net zero emissions by 2050.

All Parties to the UNFCCC have to be part of a new global agreement. Otherwise the 2-degree goal cannot be met. The level of ambitions however, will have to differ among countries. Further, countries must have the opportunity to contribute in different ways. Rich countries, and countries with high and/or increasing emissions, are expected to deliver the most. Developing countries with low emissions, on the other hand, could commit to delivering strategies and programs for development towards a low carbon economy. Or, they could be targeted to phase out existing fossil fuel subsidies. In order to secure the necessary actions in developing and fast growing economies, climate finance, technology transfer and capacity building must be provided for as part of the agreement.

It is well known that while emissions are decreasing in many of the developed countries (the US and the EU), they are increasing rapidly in fast growing economies such as those of China, India and Brazil. In relation to these countries, the issue is more the peak year of their emissions than emission reduction targets. This was reflected in the recent China-US agreement, where the US is committing to reduce emissions by 17% by 2030, whilst China shall peak emissions "as soon as possible."

For Norway it is important that the commitments in a Paris-agreement reflect the level of emissions, economy and capacity for reductions for each individual country. Hence, the division of Annex 1 and non Annex 1 countries in the Convention from 1998 must be discarded if the goals of the UNFCCC are to be met. How to deal with this change of facts, from the time of the adoption of UNFCCC up to today, is one of the most difficult issues to cope with in the period leading up to Paris.

Even if countries will have different emission reduction targets in a new agreement, it is paramount if we are to succeed that they can be added up under a common and transparent set of rules. This is a requirement for being able to measure the impact of actions taken by each country, and to see whether they are fair and comparable.

The urgent need for climate financing dominates the negotiations. Hence the actual delivery of scaled up financial support for mitigation in developing countries will be key for the outcome in Paris. In Copenhagen the figure of USD 100 billion per year by 2020 was introduced by developed countries. This amount is far from sufficient for covering all financial needs for mitigation and adaptation in developing countries. Further, the intention was never that the USD 100 billion should come entirely from public funding. A substantial part of climate financing has to be generated by private sector investments. Hence, it is important that the Paris agreement sends the necessary long term and predictable signals to the business community and investors.

A new "Green Climate Fund" (GCF) was recently established under the UNFCCC. Norway is a board member of the GCF. The GCF shall allocate funds to both mitigation and adaptation in developing countries. At the COP 20 in Lima (December 2014), the pledges to GCF reached the amount of USD 10 billion. This was the informal target for the first replenishment of the fund. Norway has pledged USD 258 million to GCF over the next four years.

Even if this target was met in Lima, the subject of climate finance will be critical in the period leading up to Paris. In fact there is a huge gap between the expectations of developing countries in this regard, and developed countries' positions. While developing countries expect to see new an additional public funding in line with the 100 billion figure introduced in Copenhagen, the developed countries argue that existing public funding and private finance made available through targeted public money must be "counted" as a part of the USD 100 billion. It will be very difficult to get any form of agreement in Paris if the issue of climate financing is not somehow settled among the Parties.

The new agreement in Paris has to secure a cost effective way of reducing emissions. It has to ensure that more will be done by concerted actions globally than the sum of measures taken individually by each country. Hence, Norway argues that Kyoto-type flexible mechanisms have to be extended and strengthened in a new agreement. It has been made quite clear in the submissions to the UNFCCC that Norway's level of ambitions in a new agreement will depend on whether these mechanisms are extended. At the current stage of the negotiations, whether or not such mechanisms will be a part of a new agreement remains undecided.

The agreement in Paris will also have to deal with many other issues in addition to mitigation. The balance between adaptation and mitigation in a new agreement is also a challenging issue. For the developing countries, and in particular the poorest and most vulnerable ones, the main focus is on adaptation, and reducing the risk for their societies due to the impact of climate change. This is understandable. Hence, all countries, and developing countries in particular, have to develop plans and measures for adaptation which are generally integrated in the overall national policy of the respective country. Even if such actions have to be supported by international climate funding, it is important that these countries also allocate national resources to adaptation. Further, measures for adaptation must not substitute national actions for reducing emissions. It is evident that both have to be done at the same time.

BRIEFLY ON NORWAY'S ENGAGEMENT IN
PROCESSES OUTSIDE THE UNFCC – THE LINK TO COP 21 IN PARIS

As mentioned above, it is now clear that an agreement in Paris would be based on a bottom-up approach. This means that realistically the sum of the figures for emission reductions in the "intended national determined contributions" (INDCs) that the Parties shall convey to the UNFCCC in the first quarter of 2015, will set the level of ambitions for 2030 in the Paris agreement. Even if this will turn out to be far from sufficient to get the global emissions on a 2-degree pathway, it is not likely that the Parties will come up with more ambitious mitigation targets between April this year and the Paris COP. Hence, the French COP Presidency, (together with the previous Peruvian, and the UN Secretary General's office), is looking into how to create a positive outcome from Paris, which not only includes the agreement under UNFCCC, but also encompasses concrete ongoing actions for emissions reductions to take place outside the auspice of the negotiations. This so-called action agenda can be linked to the UN Secretary General's climate summit in September 2014, where a number of such "actions" were launched, a substantial part of these in collaboration with industry and private sector representatives.

Norway is already engaged in many of these initiatives. Norway's engagement in the forest and land use sector, and a range of bilateral agreements with developing countries for preserving forests, is important in this context. By getting more donor countries on board, and also raising private capital, it should be possible to get developing countries to raise their ambitions for forest preservation, thereby reducing their future emissions.

Further, Norway is currently co-chairing, together with Chile, the Climate and Clean Air Coalition, (CCAC). The coalition was set up after the failure of Copenhagen, and deals with the so-called short lived climate pollutants (SLCPs). The coalition has today 100 partners, from around 60 countries. SLCPs are gases and particles with a lifetime of up to 15 years, and with a very high global warming potential. In recent years, the climate, health and environmental benefits of reducing emissions of short-lived climate forcers have received increasing international and national attention. Norway is, in collaboration with France, elaborating how efforts by the CCAC can be a part of the outcome in Paris.

At this stage it is uncertain what sort of outcome we will have in Paris. A thinkable outcome could be an agreement, applicable to all Parties, where some parts are legally binding, whilst other parts, like the mitigation ambitions for each country, do not have a legally binding status. Further, in addition to the agreement, new partnerships, bilateral and global, private and public, including ministerial decisions, could be announced, heralding new and bold initiatives for emission reductions, beyond the level of ambition in the Paris agreement. If we were to achieve something along these lines, Paris would not be a failure. There is still hope!

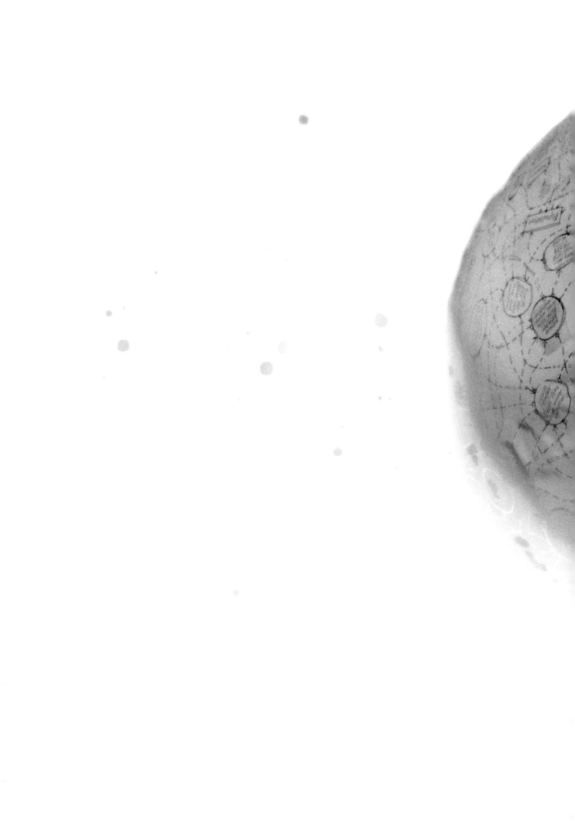

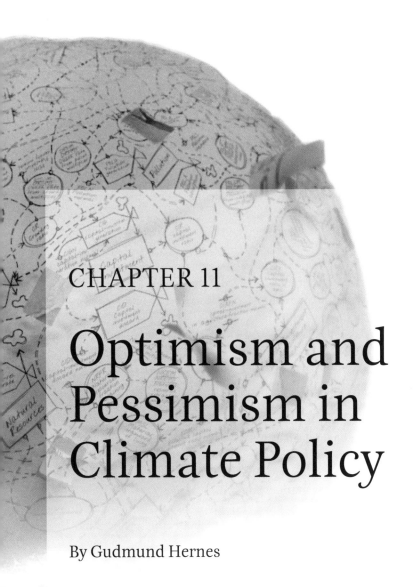

CHAPTER 11

Optimism and Pessimism in Climate Policy

By Gudmund Hernes

INTRODUCTION: SWINGS IN MOOD

In the 70 years since the Second World War, there have been wide swings in the public mood about the prospects for humanity. The war itself ended in the euphoria of V-day. The upbeat frame of mind was further stimulated by the establishment of new international institutions, such as the United Nations, The World Bank and The International Monetary Fund. There was also a sense that human interventions could better the human condition. In 1947 Germany was hit by what was called the Hunger Winter. Other countries as well were in a dismal state. But a few months later, on June 5, 1947 in a speech at Harvard, the American Secretary of State George Marshall launched the Marshall Plan for urgent rebuilding of Europe. Help was on the way, hope was rekindled! Contributing to the same lifting of spirits was the adoption of The Universal Declaration of Human Rights, in December 1948.

However, there was never just one mood – conflicting sentiments overlapped.

Another, menacing undercurrent was already becoming more manifest. It was first articulated by Winston Churchill in his Fulton Speech on March 5, 1946, when he warned against the "iron curtain descending on Europe." The communist coup in Czechoslovakia in February 1948 was a watershed. In response, on April 4, 1949, NATO was formally established. Indeed, the Marshall Plan was also in part a way to bolster people against the menace of communism. A year later, a war erupted on another continent, in Korea on June 25, 1950. The cold war was on, with hot wars in some countries! An arms race, both conventional and nuclear, was also ongoing. The threat of nuclear war became increasingly pronounced, first highlighted by the bombings of Hiroshima and Nagasaki. Then the Soviet Union tested its first A-bomb in 1949, USA its first H-bomb in 1952, followed by its Soviet counterpart the year after. Other powers established themselves as members of "the nuclear club": Great Britain (1956), China (1967) and France (1968). Now nuclear proliferation was on, as was the insidious danger from nuclear testing, where the fallout spread by both atmospheric winds and ocean currents became an issue of mounting concern.

Then, around 1960, other changes made for optimism. Not only was leadership around the world passed on to a new generation of leaders. A new cohort of nations emerged from the old colonies. And novel terms were coined; not just "the third world" but also the optimism encapsulated in the concept "developing countries." "British India" was the harbinger of what was to come, gaining its independence in 1947 as simply "India." But the biggest wave of decolonization started with Anglo-Egyptian Sudan in 1955, Tunisia from France and Morocco from Spain in 1956, Ghana, Malaya, Guinea and Iraq in 1957, followed by the rapid acceleration of liberation and independence in the early 1960s.

Decolonization manifested itself in the growth of member states of the UN as well, from the original 51 signatories in 1945 to 99 in 1960, 127 in 1970, 154 in 1980 and 189 in 2000. And the 1960s saw the establishment of the US Peace Corps and similar help organizations in many countries. "Development aid" became a fixture in the budgets of rich countries, and in a large number of programs under the auspices of organizations such as the UN and the World Bank. Such events and initiatives were signs of optimism: humans could build a better world by foresight, cooperation, education, technological change and freer trade. Indeed, applied science and good will among peoples could shape a better world and create a brighter future. A case in point was "the Green Revolution" – based on technology transfer, improved irrigation, hybrid seeds and high-yielding crops along with better infrastructure and management – that markedly increased agricultural production in the late 1960s and saved millions from starvation and gave increased life expectancy.

11.1 A NEW WAVE OF PESSIMISM

However, at the end of the 1960s, a new wave of pessimism became increasingly apparent. It was a variation on the Malthusian theme: In spite of the Green Revolution, population would outstrip resources. Three intellectual contributions can illustrate this new wave – two from 1968, and a third from 1972, of which Jorgen Randers was one of the authors.

A book that caused an enormous stir was authored by Paul R. Ehrlich (1968) and his wife Anne, entitled *The Population Bomb,*[1] warning of the mass starvation that would result from unconstrained human propagation. The alarmist opening did not mince words:

> *The battle to feed all of humanity is over. In the 1970s hundreds of millions of people will starve to death in spite of any crash programs embarked upon now. At this late date nothing can prevent a substantial increase in the world death rate. (Elrich 1968)*

The book not only provided a dark diagnosis – it also advocated an exacting program of action. Among the components for US domestic policies were:

▌ Reduce the population growth rate to zero or even negative.

---

1   Paul R. Erlich, *The Population Bomb* (New York: A Sierra Club-Ballantine book, 1968). There were earlier warnings, two dating back to 1948, Henry Fairfield Osborn, Jr. *Our Plundered Planet* (New York: Doubleday, 1948) and William Vogt, *Road to Survival* (New York: William Sloane Associates Inc., 1948) but they did no stir up the same commotion.

⟩ Greatly augment food production.
⟩ The United States must take the lead, e.g. by adding "temporary sterilants to water and staple food."
⟩ A tax system punitive for having several children.
⟩ Incentives for voluntary sterilization.
⟩ Setting up a Department of Population and Environment "with the power to take whatever steps necessary to establish a reasonable population size in the United States and put an end to the steady deterioration of our environment" (Erlich 1968, p. 138).

Among the recommendations for foreign policy were:

⟩ Rank countries by their capacity for feeding themselves and provide foreign aid only to countries that would limit population growth and be able to feed themselves in the future.
⟩ Programs for education and agricultural improvement in developing countries.
⟩ Support of more population progressive separatist movements.
⟩ Setting up selective programs for education and agriculture programs outside a more universalistic United Nations.

The book was not only alarmist – the measures proposed were draconian. And clearly they were very far from being politically feasible.

But the types of interventions were interesting: The key was that prescient central authorities must act on scientific knowledge to constrain the behavior of citizens, though Ehrlich also had suggestions for what informed citizens themselves could do, or what enlightened public opinion could do to pressure politicians to act.

The book was translated into many languages, sold more than two million copies, and alerted public opinion on the interrelation between population growth and environmental issues. It also had an impact on public policies in the ensuing years, though Ehrlich's dismal predictions overshot actual developments.

Also in 1968 the ecologist Garret Hardin published a very influential article in *Science*, titled "The Tragedy of the Commons." According to Google Scholar, by the fall of 2014 the article had been cited close to 25 000 times.

Hardin confronted the dominant tendency in thinking going back to Adam Smith to assume that "decisions reached individually, will in fact, be the best decisions for an entire society." His counter-thesis was that actors, who each act rationally in their own interest, could behave in ways which undermine the long-term common interests of them all. The result could be either excessive or damaging use of common resources. Examples ranged from overfishing and overgrazing on the one hand, to the many small emissions that result in destructive pollution. Hardin argued that such problems first of all did not have technical solutions; rather, they posed moral quandaries. Secondly, he did not believe they could be solved by personal restraint. Indeed, individual freedom combined with self-interested rational action could generate counter-productive outcomes. For example, freedom of the seas to catch whatever one sought, or freedom of entry to land, can lead to the tragedy of the commons.

Hardin saw a two-part solution to the predicament. The first was cognitive – a mental recognition that one is caught in a dilemma when everyone is hurt by everyone copying each other's rational behavior. "Individuals locked into the logic of the commons are free only to bring on universal ruin."

But from this insight flows the understanding that some other social arrangement is needed, indeed some kind of imposed restraint that can compel individuals to act in their own common interest. "[O]nce they see the necessity of mutual coercion, they become free to pursue other goals." Hence Hardin quotes with approval the Hegel/Engels dictum "Freedom is the recognition of necessity." In Hardin's catchphrase, what is required is "Mutual coercion mutually agreed upon."

Though Hardin addressed the general problem of mutually destructive behavior in the commons, his main substantive concern was, like Ehrlich, the threat of unlimited population growth. Hardin's (1968) concluding line was:

> *The only way we can preserve and nurture other and more precious freedoms is by relinquishing the freedom to breed and that very soon. 'Freedom is the recognition of necessity'– and it is the role of education to reveal to all the necessity of abandoning the freedom to breed. Only so, can we put an end to this aspect of the tragedy of the commons. (p. 1248)*

Hardin did not advocate specific policies as Ehrlich had done. Rather he analyzed a particular logic of collective action. If we put it in game theoretical terms, he addressed the problem of how a "prisoner's dilemma" could be transcended. And it is interesting that the logic of Hardin's argument closely mirrored the argument in a book published 300 years before, namely in Thomas Hobbes' *Leviathan* (1962, p. 100).

What Hardin called the "tragedy of the commons," Hobbes called "the state of nature." This was a state in which every man had the freedom to do whatever he pleased, and hence, in the famous quote, "where every man is the enemy of every other man ... and the life of man, solitary, poor, nasty, brutish, and short." And Hobbes argued that from this dismal state the insight was generated that everyone would be better off relinquishing the right to all things, and by renouncing this right, would improve their own and everyone else's well-being. But to do so would require a "common power to keep them all in awe" established by a mutual covenant (Hobbes 1962, p. 132). Or, in Hardin's compact phrase: "Mutual coercion mutually agreed upon" (Hernes 1993).

So Hardin's diagnosis was part of the general pessimism at the time – the neo-Malthusian distress that population growth might outstrip available resources in general, and agricultural carrying capacity in particular. In broad terms his was an institutional perspective on how the problem could be transcended: Not by a top-down imposition from above, like Ehrlich, but by a bottom-up perspective: actors gain insight from their mutually self-destructive behavior, and thereby see the need for self-imposed mutual constraints. However, he was not very specific about the kinds of actors he had in mind, such as whether they included governments in addition to individuals.

11.2 PUBLICATION OF *LIMITS TO GROWTH*

The third scientific contribution, which also caused a great stir in public debate and activism, was a book written by Donella H. Meadows, Dennis L. Meadows, Jorgen Randers and William W. Behrens III: *Limits to Growth* was published under the auspices of the Club of Rome in 1972. Again using Google Scholar as a measuring stick, it has been quoted nearly 12 500 times, and yet more references could be added for translations, as well as for references to the later revisits and renewed analyses published in the following decades.[2]

Limits to Growth is a much more complex analysis than the contributions of Ehrlich (1968) and Hardin (1968), both of whom basically focused on the relationship between population growth and available resources. *Limits to Growth* presented a dynamic model simulating the interactions between five key quantities: population, industrialization, pollution, food production and resource depletion. Moreover, the model included a broader set of specific variables, and incorporated nonlinearities and feedbacks in a "world model." From this the authors calculated several scenarios, and although most of them

---

2   See for example Donella Meadows, Dennis Meadows and Jorgen Randers *Beyond the Limits* (White River Junction,Vt.: Chelsea Green Pub, 1992), *Limits to Growth: The 30-Year Update*, and Jorgen Randers, *2052: A Global Forecast for the Next Forty Years* (White River Junction,Vt.: Chelsea Green Pub).

were pessimistic, the book was less of a doomsday alarm than Ehrlich's (1968) *The Population Bomb.*

Limits to Growth argued that "the relationship between the earth's limits and man's activities is changing,", "yet man does not seem to learn by running into the earth's limits." *Limits to Growth* used the whaling industry as an illustration of how an attempt to grow forever would lead to the extinction of whales and whalers. A political response could be to impose a limit on the number of whales caught each year to maintain a steady state level. Both mechanisms could be reproduced on a global scale: by noticing and then recognizing a growing danger, political countermeasures could be taken. In one sense this was in the spirit of Hardin's article.

However, delaying the choice would mean the erosion of alternatives. At some point the possibility of human determination of limits would be gone, though technological innovations could delay the reckoning. Then limits would be imposed by the brute force of circumstance, not by human choice. Human beings could recognize the relentless danger, but too late, and hence be ambushed by events.

Limits to Growth argued in favor of a *social* equilibrium model. However, where Ehrlich (1968) in *The Population Bomb*, for example, had proposed "temporary sterilants to water and staple food," the authors of *Limits to Growth* discussed politics in rather vague and abstract terms (Meadows, Meadows, Randers and Behrens 1972). Indeed, they wrote of suspending "the requirement of political feasibility and use the model to test the physical, if not the social, implications of limiting population growth" (Meadows, Meadows, Randers and Behrens 1972, p. 159). They wrote about reducing population growth in general, such as by contraception and a two child policy, and by limiting the use of resources. But they admitted that system stability could not be suddenly introduced in 1975 – stability must be approached gradually (Meadows, Meadows, Randers and Behrens 1972, p. 167). Moreover, they admitted that in contrast to modeling the interaction of population and other key resource quantities, policies must be based on mental models, as there are no formal models of social conditions and policy choice (Meadows, Meadows, Randers and Behrens 1972, p. 174).

So although controlled growth was identified as a tremendous challenge, they cast the problem primarily in terms of human values, not in terms of new institutional arrangements:

The final, most elusive, and most important information we need deals with human values. As soon as a society recognizes that it cannot maximize everything for everyone, it must begin to make choices. Should there be more people or more wealth, more wilderness or more automobiles, more food for the poor or more services for the rich? Establishing the societal answers to questions like these and translating those answers into policy is the essence of the political process. (Meadows, Meadows, Randers and Behrens 1972, p. 181 f.)

Such choices are made every day, yet few ask themselves what their choices would be, or how to take into account not only the earth itself, but also future generations. Hence, better mechanisms are needed than those of today for clarifying realistic alternatives, establishing societal goals and making short term choices consistent with long term goals. The book ends with a paragraph with dire predictions yet tempered with some hope:

If there is cause for deep concern, there is also cause for hope. Deliberately limiting growth would be difficult, but not impossible. The way to proceed is clear, and the necessary steps, although they are new ones for human society, are well within human capabilities. Man possesses, for a small moment in his history, the most powerful combination of knowledge, tools, and resources the world has ever known. He has all that is physically necessary to create a totally new form of human society – one that would be built to last for generations. The two missing ingredients are a realistic, long-term goal that can guide mankind to the equilibrium society and the human will to achieve that goal. Without such a goal and a commitment to it, short-term concerns will generate the exponential growth that drives the world system toward the limits of the earth and ultimate collapse. With that goal and that commitment, mankind would be ready now to begin a controlled, orderly transition from growth to global equilibrium. (Meadows, Meadows, Randers and Behrens 1972, p. 183 f.).

So the finale is high on moral appeal, but low on political specifics, high on emphasizing the likely future dangers but low on institutional design. Neither does the book formulate an explicit principle for institutional design, like the Hobbes/Hardin dictum of "mutual coercion mutually agreed upon."

The publication of *Limits to Growth*, with Jorgen Randers as one of the key co-authors, thus falls into the pessimistic part of the post-war oscillation between pessimism and optimism. The book had an enormous impact on the public mind, not least because of its catchy title. *Limits to Growth* encapsulated the counter-thesis to buoyant modernization theories of the previous decade which had legitimized and driven much of policy. This buoyant optimism was illustrated for example by W.W. Rostow's book from 1962 with the uplifting title *The Stages of Economic Growth*, which expected developed economies to continue growing and developing economies to move towards a "take-off."

However, *Limits to Growth* was met by a major push-back from economists in response to its main thesis. The most prominent example is probably William D. Nordhaus' article "The Allocation of Energy Resources," published in 1973, where he wrote:

> *Given the dependence on energy, there has been perennial anxiety over the adequacy of the nation's resources for meeting its apparently insatiable appetite for energy. More recently, the concern for adequacy of energy has been embedded in a more general pessimism about the viability of economic growth on a finite world. This new and pessimistic view about economic growth holds that growth is limited by a finite amount of essential, depletable natural resources. In the process of consuming finite resources, the world standard of living descends inexorably toward that of Neanderthal man. (p. 523)*

A central part of Nordhaus' (1973) argument was that the models used in works like *Limits to Growth* or and Jay W. Forrester's *World Dynamics* (1971) did not encompass the key components included in macro-economic models of growth, such as the price effects of scarcity in limiting consumption and fostering innovation, or that when prices do not take into account large external costs, this could be compensated by government imposed taxes or other constraints. On such effects, Nordhaus (1971) stated:

> *If these conclusions are right, then the current 'energy crisis' will blow over eventually. Real enough problems remain. Until supplies are expanded, the United States may experience very serious shortages or very high prices. In any case rising prices are likely over the long haul, especially for transportation; adaptation to new, potentially difficult, technologies will present a problem; and several lean years on foreign exchange markets loom ahead. But we should not be haunted by the specter of the affluent society grinding to a halt for lack of energy resources. (p. 570)*

Demographers have also addressed why world population did not result in hunger or destruction on the level predicted (Lam 2011).[3]

In sum: After the Second World War, there was a braid of optimism interlaced with pessimism over a wide range of policy issues. The ups and downs have continued until the present day. It suffices to mention on the optimistic side events such as the fall of the wall in 1989 or a book like Fukuyama's

---

3   See David Lam, "How the World Survived the Population Bomb: Lessons From 50 Years of Extraordinary Demographic History," *Demography*, 2011 (Volume 48), pp. 1231–62, which assesses the economic and demographic explanations for the surprising successes.

(1992) *The End of History and the Last Man*[4] which argued that countries and constitutions would converge towards liberal democracy as the final form of human government. On the other, pessimistic side, there are events such as 9/11 and the reports of the Intergovernmental Panel on Climate Change.

11.3 ATTITUDE CHANGE AND INSTITUTIONAL CHANGE

Limits to Growth had a great impact on public opinion: The agenda was reset and the arguments took a new direction. Yet the public attitudes about both the environment and climate have proven hard to change. In spite of dire predictions, the optimists seem to have held their ground – concern for the environmental impact of human activities and population growth in even fairly enlightened populations has declined. In the last decades there have been swings in public opinion (Hernes 2012). For example, in Norway, the overall percentage of the population who thinks pollution and environmental problems are serious, declined considerably between 1989 and 2009, though there was a slight increase after 2005. This resistance to change can be explained by what is called "the double embedding of attitudes" (Hernes 2012). The ideas people have are interlinked in a mindset, as in a logical lattice. This functions as a filter – perception is selection. So you search for what confirms what you already believe, forget what contradicts your preconceptions and ignore what jars. Psychologists have technical terms for such phenomena: confirmation bias, selective perception, selective memory, cognitive dissonance. But there is more to the story: Just as your ideas are not singletons, your beliefs are not something you cultivate in isolation from significant others. They are exchanged, shared, modified and anchored in a social network. So opinions come in ensembles, friends come in clusters – and both come together. Hence people can also support each other's mistaken beliefs.

What it then takes to change a mindset is often not the force of arguments, but the force of circumstance. For knowing more about an issue also generally means knowing more about counterarguments. Hence broad attitude change is often event-driven – i.e. happenings that are so powerful that they simultaneously shatter both the logical lattice and the social network, in such a way that both can be reconfigured. The dots can be re-connected and networks reconstructed.

---

4   Francis Fukuyama, The End of History and the Last Man (New York: Avon Books, 1992). Fukuyama has, however, turned pessimistic; see his *Political Order and Political Decay* (New York: Farrar, Straus and Giroux, 2014).

This is why the reactions to important books – through partly important events, such as the publication of *Limits to Growth* – rather than triggering a broad-based shift of opinion in the general public, often has led to the development of schools of thought.[5] The different schools become not only well versed in the arguments of their own camp, but also in how to oppose, counter and refute the arguments of their opponents. Their attitudes are doubly embedded and the contestants do not have open minds because their heads, so to speak, are densely packed with shared beliefs. So when those who were convinced by *Limits to Growth* met with, say, adherents of William Nordhaus, "the twain would never meet," so to speak.

SHARED OR CONFLICTING BELIEFS ON CLIMATE STRATEGY

One can see something of the same in schools of thought about how to counteract climate change. On the one hand you have those who in a sense build on Russell Hardin's dictum that what is needed is "mutual coercion mutually arrived at" among nations. And there are some fairly good examples of this strategy succeeding.

A good illustration of both the tragedy of the commons and institutional innovation is the curtailing of the use CFC gasses as solvents, refrigerants, and propellants in aerosol cans. It turned out that the emission of CFCs led to the depletion of the ozone layer in the upper atmosphere which protects against ultraviolet radiation. Very few who used spray cans when shaving in the morning or styling their hair knew about the gas or the aggregate effects of its use.

The question was how such issues were to be addressed. The problem was that, in addition to being unrecognized outside the circle of experts, negative effects are spread far beyond the area where they are generated, and over a much longer time horizon than that considered by the perpetrators. For example, spray cans propelled by CFC gas were primarily used in industrialized societies, but the reduction of the ozone layer, though primarily manifest in polar areas, affected people all around the globe – also those who had never used a spray can.

These negative impacts of the use of CFC were first registered in the 1970s. Then the necessity for restricting its use was addressed. About a decade later, an international agreement was reached on the reduction of the use of CFC gasses in *The Vienna Convention for the Protection of the Ozone Layer* from

---

5   For one illustration, see D.B.Luten, "The Limits-to-Growth Controversy" in K. A. Hammond, G. Macinko and W. Fairchild (eds.) (1978). Sourcebook on the Environment. (Chicago: University of Chicago Press), pp. 163–180.

1985, and under the *Montreal Protocol* from 1987 CFC was gradually phased out. (The protocol has since been revised several times.) Kofi Annan has called it "perhaps the single most successful international agreement to date" (Theozonehole.com 1987). It has been ratified by all of 197 United Nation members. The ozone hole over Antarctica has slowly been replenished, and in 2014 it was communicated that the ozone layer is recovering (The Guardian 2014). In a sense one can say that here the international community ratified "mutual coercion mutually arrived at." And it could generate a sense of optimism.

However, there are also prominent examples of international conventions that have had a mixed reception. The Kyoto protocol from 1997 is a case in point. Neither China nor the United States have ratified it, and in 2011 Canada left the Kyoto Protocol on the argument that the former two countries were not parties to it. So far the Kyoto Protocol has been ratified by only 55 countries.

The 2009 United Nations Climate Change Conference in Copenhagen also delivered fairly poor results: An accord was submitted, but not adopted, only "taken note of". The accord was not passed unanimously and it is not binding. Afterwards there was an international blame game about whose fault it was. These dismal outcomes could generate a sense of pessimism.

So what does it take to reach an international agreement? It is probably fair to say that it is easier to arrive at general principles for what should be obtained than to reach concrete and binding agreements which are implemented. The global environment is a "commons," and "mutual coercion mutually arrived" should encompass explicit goals, specific actions, definite timetables, general participation as well as clear institutional arrangements, exact reporting and clear compliance provisions. Other elements address incentives, non-participation and non-compliance penalties.[6] The problem is that reaching such agreements involves enormous transaction costs, and, as shown by the Copenhagen summit, considerable risk of failure. Hence, the question becomes: Is there another way? Do we have the time to wait for effective international agreements, particularly when "G2" – the USA and China – are leaders of opposite camps and have strong domestic groups resistant to such agreements.

---

6   See Climate Change 2007: Working Group III: Mitigation of Climate Change, http://www.ipcc. ch/publications_and_data/ar4/wg3/en/tssts-ts-13-4-international-agreements.html

11.4 COLLECTIVE ACTION OR THE
AGGREGATE EFFECTS OF INDIVIDUAL ACTIONS

In 2015 there is to be a new climate change summit in Paris. In preparation, in October 2014 a "World Summit for the Regions on Climate" was held in the French capital. Here an initiative was taken to mobilize business sectors in a bottom-up initiative. This is in line with what is called a "pledge and review initiative" by the US. At the summit, countries would put on the table the reductions in emissions each is prepared to make and to be reviewed at a later date. Some, like the former French Minister of the Environment Serge Lepeltier (2014), have come out against this strategy, arguing that "a bad climate deal is better than no climate deal":

[W]ithout agreed minimum ambitions to curb man-made global warming in 2015, the bottom-up approach could be 'an excuse' for the lack of a comprehensive effort, with scattered results. There has to be a global agreement with binding constraints. Without those commitments, what is done by local authorities and companies will remain marginal.

However, there are several examples of failure for an "all or nothing" strategy. The United Nations Climate Conference in Cancun in December 2010 made some progress towards a Green Climate Fund, but failed on carbon dioxide emissions. And when no agreement is reached, why bother to do what others will not comply with?

So there are errors of two kinds: Pursuing a bottom-up strategy that leads only to scattered results, or a collective "something is better than nothing" strategy which also leads to minuscule, though joint action. So is there another way?

Both China and the United States have been reluctant to join global conventions on climate and the environment. At the same time, both of them have taken some vigorous measures even in the absence of an agreement – also at the sub-national level.

For example, over the last decades China has undergone a massive industrialization, resulting in great damage to the environment – such as soil contamination, waste, river deterioration, deforestation, local pollution as well as global carbon emissions. The Chinese Ministry of Health has taken note: 500 million are without clean drinking water, ambient pollution kills thousands, industrial pollution has made cancer the leading cause of death, etc.[7] In

---

7   For a quick overview, see http://en.wikipedia.org/wiki/Pollution_in_China

order to counteract such hazards, various countermeasures have been taken and large-scale investments made to reduce ecological damage. Examples are:

▶ Authorities in Beijing are curbing air pollution because of the severe health and welfare costs. Put differently: Even if the Chinese authorities can afford not to reach agreements with their counterparties in international fora, in the long run they have to attend to their own population, e.g., the proportion of children afflicted by respiratory diseases in Beijing.

▶ In August 2014 it was reported that 70 smaller and medium sized cities in China will cut their production volume in order to better protect and preserve the environment and reduce poverty (Financial Times 2004).

▶ Several large-scale environment-friendly programs have been introduced. One example is the huge program to build high speed railroads (HRS) throughout the country – it is already the largest such network in the world. Another is the investment in wind power. Already in 2010 it was the largest wind energy provider in the world; in 2012 more electricity was generated from this source than from China's nuclear power stations. Wind power is identified as a key growth component in the economy.[8] Wind power also reduces the country's dependence on oil imports.

▶ Few believe that the Paris Summit on Climate in Paris in 2015 will result in a binding global cap-and-trade regime. But China has introduced large pilot projects in seven regions, and is considering a national quota system from 2016. This is important for two reasons: China has the largest emissions of climate gases in the world, and if China moves first, it could be easier for Europe and the US to follow suit. In the US, trade in CO_2-quotas is in place in several states (Alstadheim 2014).

Similarly, the United States has taken important domestic measures, also at the state, local and company level:

▶ President Obama has been working to reach a broad-reaching international agreement to compel nations to cut fossil fuel emissions that produce global warming. However, his efforts have not been successful, since a legally binding treaty requires a two-thirds majority in the US Senate. To work around this impediment, he has tried to reach a deal whereby one would "name and shame" countries into cutting emissions. This may be the only

---

8   See http://en.wikipedia.org/wiki/Wind_power_in_China

workable way given the politically gridlocked Senate which would not ratify a treaty (Davenport 2014a).

▶ In October 2014 The Pentagon released a report, *2014 Climate Change Adaptation Roadmap* (Davenport 2014b), asserting that climate change is becoming a "threat multiplier" that poses an immediate menace to national security. This threat from global warming is mediated by food and water shortages, pandemic disease and disputes over refugees and resources, and risks from terrorism. It would also increase the demand for military responses to disasters, as extreme weather triggers more humanitarian crises.

▶ The State of California is reducing emissions and energy use through more efficient state building design and construction, more renewable energy at the state level, sustainable state-owned vehicle policies, and environmentally preferable state procurement. Due to its large agricultural sector and the drought in recent years, stricter water conservation measures have been introduced.

▶ The Environmental Protection Agency has set standards for vehicle emissions. However California has adopted its own vehicle emission standards that are stricter than EPA's. Since the state has one of the largest populations in the United States, car producers have to meet California's standards if they are not to lock themselves out of this large market. But then car-buyers in other states, who also may want to sell a used car in California, find it profitable to buy one which is compliant with California rules. Often the EPA adopts California standards some years later. California's gasoline is also "cleaner" than most sold in other parts of the United States.

▶ In September 2014 Mayor de Blasio committed New York City to 80 percent reduction in greenhouse gas emissions by 2050 by overhauling the energy-efficiency standards of all its public buildings and pressuring private landlords to make similar improvements (Flegenheimer 2014a). New York will become the largest city in the world to make such a commitment, and would probably set a standard for other cities not only in the USA, but around the world.

Such more or less unilateral, local and regional initiatives are not confined to China and the United States. Carbon trading is being considered in South Korea, Indonesia, Thailand, Vietnam and Brazil. Germany has introduced its radical *Energiewende* to make renewable energy its dominant source with the final goal of doing away with coal and oil. Other countries, such as Denmark, Austria and Japan have taken steps in the same direction. Most significantly, the European Union leaders after intense negotiations in October 2014 agreed to long-term targets on climate and energy change – e.g., cutting carbon emissions by at least 40 percent by 2030 – which they hope and expect will set both ambitions and the tone ahead of the Paris climate conference in 2015 (Mock date N/A).

On the other hand, a Norwegian proposal to measure voluntary cutbacks by the same measuring stick to assess whether one approaches the 2 degrees Celsius climate goal, has so far been pushed off the table in international negotiations (see also Bjurstrom's Chapter 10; Mathismoen 2014).

In global conferences there are not just the problems of reaching a deal, but also the additional problem of agreeing on principles. The latter is often more onerous and more consequential, with repercussions not just for climate and the environment.

Hence the question is whether there is another way. Do we have to wait for effective global treaties?

In November 12, 2014 we got a first, important answer. The US President Obama and Chinese President Xi Jinping announced a joint climate commitment. The US pledged to cut greenhouse gas emissions by 26–28 percent by 2025 compared to 2005 levels, while China pledged to peak such gas emissions by 2030, sooner if possible, and to get 20 percent of its energy from non-fossil sources by that time (The White House 2014). This may have broken the stalemate that has repeatedly stymied international climate talks. Other big producers of climate gasses, notably India, may follow their lead. A key reason is the domestic air pollution and the billions of years of lost life expectancy. Citizens have called for change locally (Greenstore 2014).

CONCLUSION: LIMITS TO WHAT?

So where do we stand with respect to the dire predictions of *The Population Bomb* and *Limits to Growth*, or with respect to the dictum by Russell Hardin that agreements to overcome the tragedy of the commons must be based on "mutual coercion mutually arrived at"? Should we, so to speak, err on the side of optimism or pessimism?

Even though the predictions in *Limits to Growth* were dire, the authors were overly pessimistic with respect to innovative future technologies – e.g., on the supply of oil partly by new discoveries, and most recently by fracking, but also with respect to fuel efficiency, building methods for the growing use of electric cars or solar and wind power. On the other hand, the impacts of global warming were underestimated, particularly the social repercussions such as those reviewed in the latest Pentagon report cited above.

In a sense, *Limits to Growth* was a harbinger of the view that the genie is out of the bottle – the powers that science has given humans by careless use also threaten to cause great damage to humanity. This provides an important lesson about the link between the subject matter of science and social science. *Limits to Growth* was seminal because it helped us develop a language for what was happening.

Since the 1970s, *Limits to Growth* has had its sequels, and The UN International Panel on Climate Change (IPCC) has published its groundbreaking reports. More than that, Mother Nature herself is sending a loud, strong and clear message about what is happening to our planet. Scientists tell us that temperatures are rising, icecaps are melting, oceans are swelling, and weather is becoming more violent and unpredictable. These effects are inescapable, they are relentless – and they are our common destiny: a challenge to all humans and our commons – and there is no escape, no Planet B in sight.

Both *Limits to Growth* and The Climate Panel have taught us one more thing: The changes we see are not the forces of nature autonomously at work, like planetary motions. The processes we observe are surely geophysical, chemical, and meteorological. And they translate into processes that are ecological and biological – all species will be affected.

But what has set the forces of nature in motion is human action – and inaction! The key causes of climate change are primarily social. And for the human species the grave consequences will also be social. Land for agriculture will be destroyed by inundations, pollutants, soil loss and desertification. Clean water and food will be in shorter supply. Diseases will spread. Social inequality will be sharpened. Migration will mount from climate change refugees. Extreme weather and social crises can multiply and conflicts may be provoked.

In sum: All the social problems we face will be amplified and our common predicaments exacerbated. The poor will be hurt the most; those with the least resources will face the gravest impacts.

We cannot change the way the forces of nature work – but we can change the ways in which humans interact with them. These responses must be political – collective and binding when we can, as well as individual and voluntary when we must. Or even hybrid when possible.

This is why social science is so critical for the destiny of our planet afflicted by climate change, in identifying the social causes of climate change, mapping human impacts, calculating costs and advising policies. Social science based activism can help measure, assess, negotiate and organize – and of course also help preserve human diversity and culture. Predictions about climate change can make us pessimistic. But actions which address it and at the same time, even if in a piecemeal fashion, bring forth a better world, can make us more optimistic.

BIBLIOGRAPHY

Alstadheim, K.B. (2014) CO2-pris? Hvilken, *Dagens Næringsliv* 4.9.2014.

Davenport, C. (2014a, 26 August) Obama Pursuing Climate Accord in Lieu of Treaty, *Nytimes.com (Online),* http://www.nytimes.com/2014/08/27/us/politics/obama-pursuing-climate-accord-in-lieu-of-treaty.html?_r=0

Davenport, C. (2014b, 10 October) Pentagon Signals Security Risks of Climate Change, *Nytimes.com (Online),* http://www.nytimes.com/2014/10/14/us/pentagon-says-global-warming-presents-immediate-security-threat.html?ref=earth

Erlich, P.R. (1968) *The Population Bomb,* New York: A Sierra Club-Ballantine book.

Financial Times (2004). Small Chinese cities steer away from GDP as a measure of success, *Financial Times,* Aug 13, 2004, http://www.ft.com/intl/cms/s/0/a0288bd4-22b0-11e4-8dae-00144feabdc0.html#axzz3AjcFRXHr

Flegenheimer, M. (2014a, 20 September). De Blasio Orders a Greener City, Setting Goals for Energy Efficiency of Buildings, *Nytimes.com,* http://www.nytimes.com/2014/09/21/nyregion/new-york-city-plans-major-energy-efficiency-improvements-in-its-buildings.html

Forrester, J.W. (1971). *World Dynamics.* Cambridge: Wright. LN.

Fukuyama, F. (1992). *The End of History and the Last Man.* New York: Avon Books.

Greenstore, M. (2014). The Next Big Climate Question: Will India Follow China, *Nytimes.com (online),* http://www.nytimes.com/2014/12/03/upshot/the-next-big-climate-question-will-india-follow-china.html?abt=0002&abg=1

Hardin, G. (1968). The Tragedy of the Commons. *Science, 162*(3859), 1243–1248

Hernes, G. (1993). Hobbes and Coleman, in A.B. Sørensen and S. Spilerman (Eds.) (1993), *Social Theory and Social Policy. Essays in Honor of James. S. Coleman,* pp. 93–106. Westport: Praeger.

Hernes, G. (2012). *Hot Topic – Cold Comfort. Climate Change and Attitude Change,* Oslo: Nordforsk, Pdf downloadable at http://www.nordforsk.org/en/publications/publications_container/hot-topic-cold-comfort-climate-change-and-attitude-change

Lam, D. (2011). How the World Survived the Population Bomb: Lessons From 50 Years of Extraordinary Demographic History, *Demography, 48,* 1231–62.

Lepeltier, S. (2014). cited in Nelson, A. (2014) No Paris climate deal better than bad one – former French climate minister, *The Guardian,* http://

www.theguardian.com/environment/2014/oct/10/no-paris-climate-deal-better-than-bad-one-former-french-climate-minister

Mathismoen, O. (2014). Ingen ny bindende klimaavtale, *Aftenposten* 1/9-2014.

Meadows, D.H., Meadows, D.L., Randers, J. and Behrens, W.W. (1972). *The Limits to Growth.* New York: Universe Books.

Mock, V. (year not available). EU Leaders Agree to Long-Term Energy, Climate Change Target, *The Wall Street Journal,* http://online.wsj.com/articles/eu-leaders-seek-to-bridge-divide-over-climate-goals-1414087982

Nordhaus, W.D. (1973). The Allocation of Energy Resources Brookings Papers on Economic, *Brookings Papers on Economic Activity,* 4(3), 529–576,

Rostow, W.W. (1962). *The Stages of Economic Growth.* London: Cambridge University Press.

The Guardian (2014) Ozone layer shows signs of recovery after 1987 ban on damaging gases, *The Guardian,* Sept 10, 2014, http://www.theguardian.com/environment/2014/sep/10/ozone-layer-recovery-report-shows

The Ozone Hole (1987). *The Montreal Protocol on Substances that Deplete the Ozone Layer. Theozonehole.com.* 16 September 1987. For a good overview, se http://en.wikipedia.org/wiki/Montreal_Protocol

The White House (2014). U.S.-China Joint Announcement on Climate Change, *Office of the Press Secretary,* Beijing, China. 12 November 2014. http://www.whitehouse.gov/the-press-office/2014/11/11/us-china-joint-announcement-climate-change

Part IV System Dynamics and Modeling

The final part of the book looks at the foundation of Jorgen's activism: modeling. During his formative years at MIT in the early 1970s his work was marked by the strong and lifelong influence of system dynamics modeling.

As soon as system dynamics modeling hit the international arena with the publication of *Limits to Growth* in 1972, economists sprung into action. The first chapter in this section, Chapter 12 by Victor Norman, discusses both the role of systems dynamics in environmental activism and the conflict between the worldviews of traditional economics and system dynamics linked to the issues of resource scarcity and economic growth. Finally, he questions how radical the contribution of system dynamics really was. Chapter 13, by MIT's leading system dynamics professor John Sterman, looks into how to use modeling for educating the general public through do-it-yourself types of simulations. Chapter 14 by Ulrich Golüke looks at an issue on which Randers has spent a lot of time and effort: using system dynamics to forecast the shipping cycles models. This section then closes with an environmental concern very close to Jorgen's heart, the future of the forests as biofuels in relation to climate: Bjart Holtsmark's Chapter 15 questions whether the increased use of biofuels from forests is a good idea.

When the issues are complex, having a good systems model is key for putting activism on a sound track. This chapter concludes the Festschrift by emphasizing the need for science based activism and the challenges of actually doing it – in a smart way with a valid model as foundation.

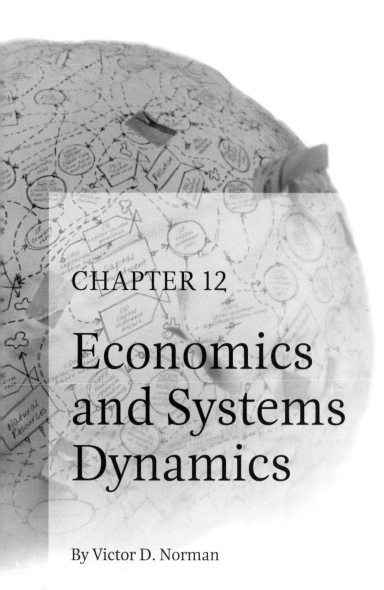

CHAPTER 12

Economics and Systems Dynamics

By Victor D. Norman

INTRODUCTION

I first met Jorgen Randers in 1970, when we were both graduate students at MIT. I was doing a Ph.D. on economic growth theory, at the Department of Economics; he was doing the same, on dynamics of social change, in the industrial dynamics group at the Sloan School of Management next door. At the same time, he co-authored *Limits to Growth* – the influential and devastating attack on the sustainability of economic growth in a world with exhaustible natural resources. We disagreed then, and we have continued to disagree on many issues since; but we have always been friends and good colleagues. While I, along with most other economists, have been critical of the approach that Jorgen and others in the system dynamics tradition have taken to resource depletion and other environmental issues, I have always admired his strong and consistent dedication to the environment and his choice of an activist approach to the task of convincing the world of the dire consequences of the status quo.

I therefore welcome this opportunity to revisit old debates – to explore old controversies and to try to assess what (if anything) we have learned since then.

When Jay Forrester published *World Dynamics* in 1971 – the first application of systems dynamics to resource scarcity and economic growth – he provoked many economists. A famous response was an article by William Nordhaus (1973) in which he argued that Forrester's modeling approach was nothing new, that his economic theory was a major step backwards, and that his simulations were an exercise in measurement without data. The success that Jorgen and his co-authors had with *Limits to Growth* – the follow-up on Forrester – did not serve to soften the criticism. I remember a colleague of mine who, in a debate with Jorgen, said that he was going to ask and answer three questions about *Limits to Growth*: Is it important? Is it good? Is it well written? His answers were no, no, and no.

Not all critics have mellowed with age. When the authors of *Limits to Growth* published a sequel in 1992, Nordhaus (1992) responded with an article named "Lethal Models 2." Nevertheless, after the immediate, virtually unanimously negative response, some economists took a more constructive view. They did not agree with the doomsday predictions of *World Dynamics* and *Limits to Growth*, but they accepted that the books raised important questions. A case in point was Robert Solow, the founder of the modern theory of economic growth, who, in his Richard Ely lecture to the American Economic Association in May, 1973, said,

"About a year ago, having … like everyone else, been suckered into reading the *Limits to Growth*, I decided I ought to find out what economic theory has to say about the problems connected with exhaustible resources" (Solow (1974)).

I shall try to follow in the Solow tradition. Specifically, I shall look at three questions. The first is what mainstream economics actually has to say about the relationship between exhaustible natural resources and economic growth. The second is how the results from mainstream economics relate to analyses in the systems dynamics tradition. The third could have been who is right, but as I am a mainstream economist by profession, my answer to that question is predictable. Instead, the last question will be what reasonable people, with open minds, could learn from a comparison between the two approaches to natural resource scarcity.

I am not going to discuss the merits or weaknesses of systems dynamics as such. It is undoubtedly a powerful tool for dynamic simulation of large systems. To ask whether one is for or against systems dynamics is as meaningless as asking whether you are for or against addition or multiplication. Systems dynamics is a valuable addition to the toolkit of engineers, business analysts, economists and others concerned with physical, ecological, social or economic dynamics. For economists in particular, it can be a valuable alternative or supplement to numerical partial and general equilibrium models of particular markets or of the economy as a whole.

My concern, however, is with the applications of systems dynamics; not with the tool. The central questions that these applications raise address the extent to which natural resource scarcity (including irreversible changes in the natural environment brought about by economic activity) places absolute constraints on economic well-being and economic growth.

12.1 NATURAL RESOURCES AND ECONOMIC GROWTH

The subject matter of economics is scarcity of resources and the consequent constraints on our ability to produce goods and services.

For the classical economists, land (or more generally natural resources) was the ultimate scarce resource and therefore in the long run the binding constraint on man's ability to produce. This led to the Malthusian pessimism: It is impossible to raise living standards above the subsistence level, because at a living standard above this level, the population will grow, and for a given supply of land this will make the marginal product of labor fall – and the standard of living with it.

Even though the Malthusian prediction proved wrong almost from the day it was formulated, no consistent, alternative theory was developed until Solow and others in the 1950s developed the so-called neoclassical theory of economic growth. The point of that theory is that economic growth is possible through accumulation of reproducible resources – capital goods of various types (production equipment and other forms of real capital, education and other forms of human capital, new technology or other forms of new knowledge). Living standards can rise so long as the accumulation of such capital exceeds the rate of population growth. The prediction of neoclassical growth theory is, therefore, that living standards will reach a fixed, steady-state level above the subsistence level. The level will be determined by the saving rate, the rate of capital depreciation, and the rate of population growth.

Neoclassical growth theory does not explicitly include natural resources. Instead, it implicitly assumes that natural resources can be regarded as a capital good – along with machines, knowledge and others. When measuring how inputs and outputs change over time, one should take care to subtract depletion of non-renewables when calculating the change in the total capital stock; but that is all you have to do.

This is the crucial stage at which neoclassical theory departs from the classical pessimism (and from *Limits to Growth*). The treatment of natural resources (and all the other different capital goods that constitute the "capital stock" in the neoclassical model) implicitly assumes that different capital goods (including non-renewable natural resources) are perfect substitutes.

Are they?

In an obvious sense, they are not. You cannot directly replace a ton of oil by a machine; or an acre of land by a figment of the imagination; or an irreversible rise in global temperature by massive investment in abatement.

In a deeper sense, however, they are – provided there are complete and perfect markets for all capital goods (including all natural resources and all forms of knowledge).

To make this clear and precise, it is necessary to use a simple, mathematical model: Suppose, as an example, that goods can be produced by a natural resource (call it copper) and a real capital good (call it machines) according to some production function

(1) $X = F(M,C)$

where X is output, M is the stock of machines and C is the amount of copper used in production. There is, obviously, no reason to expect that machines and copper are perfect substitutes at this level.

If we add profit-maximizing behavior and perfect markets to (1), however, we get something which looks very much like perfect substitution:

To make things simple, assume that there are no extraction costs for copper, and measure output and the stock of machines in the same units (real $, say). Let p denote the market price of copper and r the real interest rate. With profit-maximizing firms and perfectly competitive markets, these prices will equal the marginal products of copper and machines, respectively, so we shall have

(2) $\dfrac{\partial X}{\partial C} = p$ $\dfrac{\partial X}{\partial M} = r$

National income, properly measured, is total production less the opportunity value of the natural resources used today instead of being saved for future generations. Let us denote by v the opportunity value of copper. True national income Y is then

(3) $Y = X - pC$

With a perfect market for the natural resource, the price will equal the opportunity value – it must be equally profitable to extract copper today as to leave it in the ground for future use. It follows that the price over time must rise at the rate of interest – the so-called Hotelling rule. Thus, we have

(4) $v = p$ $\dfrac{dp}{dt} = rp$

To see what this implies, note first from (1) and (3) that growth in national income must be

(5) $\dfrac{dY}{dt} = \dfrac{\partial X}{\partial M}\dfrac{dM}{dt} + \left(\dfrac{\partial X}{\partial C} - p\right)\dfrac{dC}{dt} - C\dfrac{dp}{dt}$

The second term on the right-hand side is zero since the price of copper equals its marginal product. Using (2) and (4), we are then left with

(6) $\dfrac{dY}{dt} = r\left[\dfrac{dM}{dt} - pC\right]$

which says that growth in (true) national income is the rate of return on capital times the change in the capital stock including natural resource depletion.

The important thing to note from (6) is that investment in new machines and depletion of natural resources enter additively – it does not matter whether we invest in machines and deplete natural resources or leave the natural resources in the ground; all that matters is the difference between machine investments and resource depletion. That is the sense in which capital goods and natural resources are perfect substitutes.[1]

---

1   It is tempting to use (6) to argue that we can aggregate stocks of the different capital goods as well into a single capital aggregate. Whether that is possible or not was the subject of the so-called capital controversy in the 1960s. The general answer is no – but that it would be so nice if we could that we may be excused for not resisting the temptation.

The fact that the two are perfect substitutes in this sense, does not, however, mean that we can replace natural resources with reproducible capital without loss of productivity and growth. That depends on the degree of substitutability in the physical sense.

To see why, let us extend the simple model above to a complete growth model by assuming that the economy has a true, net saving rate of s; i.e. that total net investment in new machinery, minus the value of depleted natural resources, adds up to s percent of net income:

(7) $\quad \dfrac{dM}{dt} - pC = sY$

Substituting (7) into (6), we then see that the rate of growth of the economy will be

(8) $\quad \dfrac{dY/dt}{Y} = sr$

so, for example, if the rate of return on capital is 5 percent and the saving rate is 20 percent, the growth rate will be 1 percent.

For a given saving rate, the growth rate will therefore be determined by the rate of return on capital. In this context, r is not some return on capital in the abstract sense, however – it is the marginal productivity of machines ($\partial X / \partial M$ from equation (2)). To what extent rapid growth continues to be possible as natural resources are depleted and are being replaced by machinery, is therefore a question of how that substitution affects the productivity of machinery.

If it is easy to reduce natural resource inputs per unit of output through investment in new or better machinery, then the marginal productivity of capital will not fall significantly as natural resources are depleted. In that case, therefore, long-run growth will not be significantly affected by increased natural resource scarcity. If, alternatively, we cannot easily replace natural resources by reproducible capital, the resources will effectively limit economic growth.

The cutting point between the two is at an elasticity of substitution (the elasticity of the demand for machines relative to natural resources with respect to the relative price of the two) equal to unity. At an elasticity below unity, natural resources will be a limiting factor on production and growth; at an elasticity above unity, long-run economic growth will be possible even with exhaustible natural resources.

12.2 GROWTH THEORY VS. LIMITS TO GROWTH

How do these conclusions from mainstream growth theory relate to the absolute limits predicted by systems-dynamics analyses?

The immediate answer would seem to be that it is all about the elasticity of substitution: If you believe in a high elasticity of substitution, you can leave *Limits to Growth* unread on the shelf. If you believe that the elasticity is low, read it (and read it again).

Up to a point, that is also the correct answer. Much of the disagreement between economists and the authors of *Limits to Growth* is about substitutability. Economists typically believe in substitution. In fact, the reason the price mechanism functions, is that individuals and firms respond to price changes by substituting goods which have become more expensive with goods which have not. The evidence that prices affect behavior is overwhelming.

Moreover, if we think through the types of substitution which will be induced if prices of exhaustible natural resources rise, relative to prices of other goods and services, there is every reason to believe that the overall substitution effect is large: If, say, copper becomes more expensive relative to other inputs and goods, it will (1) make manufacturers search for other raw materials to replace copper, (2) induce those without such alternatives to use more man-hours to reduce the amount of copper going directly to waste, (3) make the firms shift to less copper-intensive modes of production (including investing in more copper-efficient machinery), (4) give incentives to spend R&D resources to develop new, less copper-intensive technology for production, and (5) cause consumers to change their consumption patterns from copper-intensive goods to other goods and services.

The sum total of these substitutions will in the majority of cases imply a substitution elasticity well above 1.

A good example, although it has nothing to do with natural resources, is given by the solution to what the upper classes in Europe in the early 1900s called "the servant problem" – the problem that arose as the unskilled wage in manufacturing rose to a level where fewer and fewer wanted to work as servants in private homes. We hear little about that problem today. It resolved itself, by the higher cost of servants, by rich people moving into smaller houses, by having fewer children, by eating simpler meals, by the development of labor-saving household appliances – and also by the acceptance of less tidy and less clean homes. Instead, they now have many, large and fast cars, travel incessantly around the globe, and eat in fashionable restaurants. The sum total has been substitution away from a labor-intensive to a more capital- and energy-intensive pattern of consumption.

The example is good because it illustrates three important points about substitution. The first is that substitution takes time – in the short run, the only thing the landed gentry in England could do about the servant problem was to make do without one or two of the footmen or maids; over a genera-tion or two, they found ways of doing without the entire staff (and the estate with it). The second point is that substitution is about much more than the technical question of how to produce a particular good using less of a natural resource – long-term substitution is typically about changes in lifestyles and the patterns of consumption, and about the corresponding changes in indus-trial structure. The third point is that innovations and new inventions are typically also a form of substitution, reflecting the importance of economic incentives in the development of new technology.

Up to a point, the controversy between economists and system-dynam-ics analysts is probably due to a mutual misunderstanding on these points. One way to read *Limits to Growth* is as a warning of what will happen unless we change lifestyles, consumption patterns and industrial structures in ways that drastically reduce the use of exhaustible natural resources. One way to interpret the criticism of the book is that it fails to take into account the possibilities to rapidly change lifestyles, consumption patterns and industrial structures.

There are, however, good reasons to believe that the disagreement goes deeper – at least between some economists and some system-dynamics ana-lysts. The more fundamental disagreement has to do with whether the econ-omy will self-correct early enough as exhaustible resources become more scarce. That has to do with whether we have a price system for exhaustible resources that provides correct incentives for substitution.

To see the issues involved here, we have to go deeper into the Hotelling rule for pricing of exhaustible resources in a perfect market.

12.3 MARKETS AND PRICES FOR EXHAUSTIBLE RESOURCES

The Hotelling rule tells us how prices of non-renewable natural resources must evolve to ensure supply of such resources at all points in time. In its general form, it says that the rent from a unit of the resource must grow at the rate of interest to give sellers incentives not to extract all of the resource today or leave all of it in the ground for the future. In a simple case of zero extraction costs, this means that the price must rise over time at the rate of interest. That tells us what the price path will look like. With a constant interest rate, it says that the path must be an exponential function

(9) $p(t) = p(0)e^{rt}$

To describe the natural resource market completely, however, we also need to know what determines the *level* of the price path – i.e. the constant term $p(0)$ in equation (9).

The answer is straightforward: The price path must be such that the stock of the natural resource is exactly sufficient to satisfy demand for it over the relevant time horizon. Using an infinite horizon, that means that $p(0)$ in equation (9) must be given by the condition that the reserves are just sufficient to satisfy total consumption, i.e.

(10) $\int_{0}^{\infty} C(t)dt = R$

where R is the stock of the resource at time 0.

It could well follow from (10) that the reserves are completely depleted in finite time. That will typically be the case for natural resources that can easily be substituted by other resources or can be made redundant by changes in consumption patterns.

For resources which cannot be fully substituted, however, the resource must last forever.

If a finite stock is to last forever, consumption of it must be infinitesimally small from some point of time onwards. It need not drop to zero (consumption could, for example, evolve like a geometric series with an infinite number of terms and still up to a finite sum), but it will have to become very close to it. That will eventually and inevitably give a limit to growth. It is, however, probably not, for most exhaustible resources, as Joseph Stiglitz points out in the quotation below, a limit that is relevant over an appropriate time horizon:

> ... *it is obvious that continued exponential growth is impossible, if only because eventually, at a strictly positive growth rate, the mass of people would exceed the mass of the earth. I am not concerned here with such very long-run problems. I am concerned here with the more immediate future. (Stiglitz 1980)*

For most of us, "the more immediate future" would not go beyond 2–300 years.

The important point about the Hotelling rule is that if prices are given by (9) and (10), producers and consumers will be given correct incentives; so the use of exhaustible resources – and their gradual phasing-out as stocks are depleted – will give an efficient allocation of natural resources over time.

If we believe that actual prices satisfy the Hotelling conditions, therefore, there should be no reason to be concerned about exhaustible resources. They might still constrain long-term economic growth, but there would be nothing useful for us to do about that, since the price mechanism would ensure that we would get maximum mileage out of what we have.

So are there reasons to believe that actual prices are Hotelling prices?

For some very important resources, we either have very imperfect markets (water is a prime example) or no markets and prices at all (carbon emissions). I shall return to these below, so for the time being, let us concentrate on those resources for which property rights are well-defined and respected, and where we therefore have well-defined markets.

As Solow (1974) pointed out, prices of exhaustible resources serve two very different functions simultaneously: They are prices for day-to-day transactions for the flow of extracted resources from producers to firms and consumers. And they are also asset prices for the resource reserves. As transactions prices their function is simply to clear the market. As asset prices, their function is to ensure that current and future prices give incentives to an optimum use of the reserves over time.

There is reason to believe that natural resource prices function reasonably well as market-clearing prices for current transactions. They typically – like copper – respond rapidly and strongly to variations in demand and supply. The markets are not always perfectly competitive. Some, like the market for oil, are very far from the competitive ideal. But there are significant checks on exploitation of market power. And to the extent that OPEC and others exert market power, it is more likely that the result is a conservationist bias in prices – i.e. too high rather than too low prices.

The same may not hold true for natural resource prices as asset prices: All asset prices – of natural resources, of property, of stocks and bonds – are far too volatile to have credibility as indicators of future value. The reason is probably, as Keynes pointed out, that we do not have a good basis for assessing future values generally:

> *The outstanding fact is the extreme precariousness of the basis of knowledge on which our estimates of prospective yield have to be made. Our knowledge of the factors which will govern the yield of an investment some years hence is usually very slight and often negligible. If we speak frankly, we have to admit that our basis of knowledge for estimating the yield ten years hence of a railway, a copper mine, a textile factory, the goodwill of a patent medicine, an Atlantic liner, a building in the City of London amounts to little and sometimes to nothing; or even five years hence. (Keynes 1936)*

In the absence of long-term knowledge, we seek refuge in shorter-term havens. The asset market is, paradoxically, in itself such a haven: To assess whether it is profitable to buy a copper mine today, it is sufficient to have a reasoned opinion about the market value of the same mine tomorrow. That value has nothing to do with the long-term fundamentals in the market. It is simply a question of how the copper market sentiment will change overnight. A good speculator, therefore, is one who is good at assessing short-term market sentiments – what Keynes called "animal spirits" – not one whose expertise is long-term assessment of market fundamentals.

If this is true for asset markets generally, it must almost certainly be true for markets for exhaustible resources, where the relevant time horizon for correct valuation is many times as long as it is for stocks, bonds or housing.

As an illustration, suppose the current market price of a natural resource is $150 per ton, and that we know that this is the correct Hotelling valuation, at a 4 percent real interest rate, if the resource deposits are sufficient to last for another 225 years. Then the corresponding correct valuation would be close to $400 if we knew that the reserves would last for only 200 years, and $50 if they would last for as long as 250 years. You would have to be more than exceptionally well-informed to have any idea as to whether the current price is too high or too low.

The controversy between economists and systems-dynamics people is, in fact, in itself the best example of our lack of the type of knowledge needed for correct, long-term pricing of exhaustible resources.

This does not necessarily mean that current prices are biased estimates of the discounted future value of natural resources. If anything, there is reason to believe that they are unbiased, as any clear indication of bias would cause current prices to rise or fall to eliminate the bias. The point is only that current prices are grossly inefficient estimates of future values – the information content about long-term scarcities is close to zero.

CONCLUSION: REASONABLE INSIGHTS

Where does this leave us when all is said (but, alas, too little is done)? Should we share the pessimism of Forrester, Randers and his disciples? Or should we place our faith (and fate) in market-induced substitution?

As far as "ordinary" natural resources are concerned, I would still go for mainstream economics. Current prices could be too low or too high to ensure the appropriate longevity of the resource – we don't know – but they will self-correct as we go along. If current prices are too high, we shall experience a future resource glut. If they are too low, we are in for some negative surprises some decades ahead. In the former case, we shall discover that we have substituted away resource-intensive goods and production methods too soon. If too high, the substitution challenge kicks in later. It will then be more pressing than if an optimum adjustment path was followed. In both cases, we will incur some unnecessary costs; but they are likely to be well below our threshold of pain.

Replacing copper wires with wires with synthetic fibers ten years too early or too late (compared to what the true reserves of copper should have us do) is something we can well live with.

Moreover, the alternative − centrally planned management of natural resources − is at least as likely to get the correct time path wrong. We should bear in mind that the problem with a market solution for exhaustible resources is not the market mechanism; the problem is the lack of precise knowledge about future scarcities. That lack of precise knowledge is shared by − or even worse for − any central planner.

Unfortunately, some important natural resources are not ordinary. Water resources are more often shared in ways that preclude rational exercise of ownership rights; the world atmosphere is treated as a free garbage dump. Because of these and similar resources, Forrester-Meadows-Randers' projections are likely to materialize unless we do something to correct the market failures.

Economists have good answers to these problems − or rather, they have *a* good answer: All you have to do is to award property rights. The market will then do the rest. If, in the Kyoto tradition, we award property rights to carbon emissions by giving fixed emission quotas to each country in the world, and allow the countries to trade quotas, we can both save the world climate and ensure an efficient distribution of emission rights between activities and countries.

A good answer is, unfortunately, not a good solution unless it is implemented. And implementation requires that policy makers and the general public have a clear perception of what the alternative is. No one should realize this better than economists, who as teachers spend most of their time trying to explain opportunity costs and choices between alternatives to students. One of the remarkable things about the response of my profession to *Limits to Growth* was that few, if any, saw the dynamic simulations as exactly that: An illustration of what would happen if proper action was not taken. They (we) may be excused when it comes to those natural resources for which there are well-functioning markets. For non-market resources like water and air, however, the profession should have welcomed the Forrester-Meadows-Randers simulations as a brilliant sales pitch for action − and an opportunity for economists to piggyback on the wave of concern in order to sell the good answer.

Fortunately, even economists learn. In the case of climate change, the Stern Commission tried in 2006 to do exactly the same as Jorgen and his colleagues did in 1972 – show the long-term consequences of inaction.

At the end of the day, we may not disagree very much, after all. Let us only hope that it happens soon enough to save the world climate.

BIBLIOGRAPHY

Forrester, J.W. (1971). *World Dynamics*. Cambridge: Wright. LN.

Keynes (1936). *The General Theory of Employment Interest and Money*. London: Macmillan.

Meadows, D.H., Meadows, D.L., Randers, J. and Behrens, W.W. (1972). *The limits to growth*. New York: Universe Books, 102.

Meadows, D.H., Meadows, D.L. and Randers, J. (1992). *Beyond the limits: global collapse or a sustainable future*. Earthscan Publications Ltd.

Nordhaus, W.D. (1973). World dynamics: measurement without data. *The Economic Journal, 83*(332), 1156–1183.

Nordhaus, W.D., Stavins, R.N. and Weitzman, M.L. (1992). Lethal model 2: The limits to growth revisited. *Brookings Papers on Economic Activity*, 1–59.

Solow, R.M. (1974). The economics of resources or the resources of economics. *The American Economic Review, 64*(2), 1–14.

Stern, N.H. (2006). *Stern Review: The economics of climate change, 30*. London: HM treasury.

Stiglitz, J.E. (1980). A neoclassical analysis of the economics of natural resources. *NBER Working Paper*, (R0077).

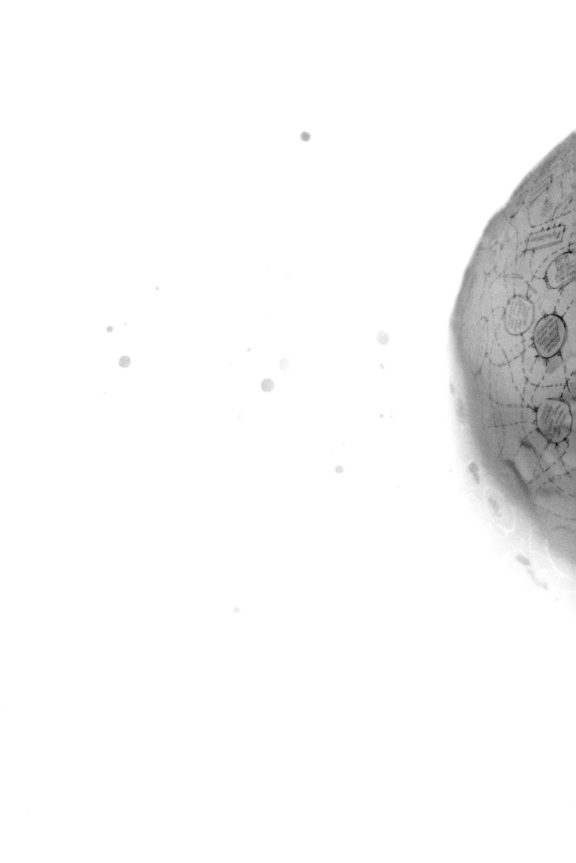

CHAPTER 13

Learning for Ourselves: Interactive Simulations to Catalyse Science Based Environmental Activism

By John D. Sterman

INTRODUCTION

Humanity has never been as healthy, rich and numerous as it is today, but at the same time, we have never faced greater and more difficult threats to our health, wealth and survival. The developed nations consume disproportionate shares of global resources and seek further economic growth, while billions lack the food, housing, healthcare, electricity, mobility, education, and other resources the affluent take for granted, and legitimately seek to meet these basic needs. However, the global environmental footprint of humanity is already an unsustainable 1.5 Earths (See Wackernagel, this volume). And it is growing: the UN projects world population will exceed 10 billion by 2100, and exponential growth in global gross product is rapidly worsening environmental problems from collapsing fisheries to water shortage to extinction of species to climate change (Meadows, Randers and Meadows 2004, Randers 2012, Rockström et al. 2009, Running 2012).

The traditional response to any environmental problem is trust in technological innovation and market responses. Resource depletion or environmental degradation, it is argued, will raise the price of any scarce resources, leading to innovation that solves the problem (see Norman, this volume). Where there are market failures that prevent the price system from functioning, as for climate change and many other environmental problems, it is the role of scientists and other experts to carry out research identifying threats and potential solutions and communicate the results to political leaders, who will then take actions to mitigate the damages through regulation or government-funded technological innovation.

That approach, which I call the Manhattan Project model, does not and cannot work for the most important environmental problems we face. Why? In 1939, faced with existential threats, Albert Einstein and other scientists personally alerted President Roosevelt of developments in atomic physics promising weapons of unimaginable power. The President listened to the experts. Then, by focusing enough money and genius in the deserts of New Mexico, the US created nuclear weapons in just six years. The appeal of the Manhattan Project model for environmental threats is clear: it is a *deus ex machina*, a technical fix that offers the prospect of solutions to tough problems without requiring sacrifice, contentious political battles or deep changes in our way of life.

But a Manhattan Project cannot solve the climate problem or other pressing environmental threats. The US developed the bomb alone, in secret, with no role for the public. In contrast, no single nation can solve the climate problem on its own or without the involvement of the public. The climate is a common-pool resource and vulnerable to the Tragedy of the Commons;

solutions require collective action at multiple scales, from individual behaviour to government actions at local, national and international levels (see Hernes, this volume). Cutting Greenhouse Gas (GHG) emissions enough to prevent additional harm from climate change requires that everyone cut their carbon footprints, requiring dramatic changes in the sources and uses of energy – and the political support to enact the policies and legislative changes needed to implement these changes in time.

For many years there has been no reasonable doubt that the climate is changing, that most of these changes are caused by human activity, and that continuing on the traditional path creates high risk of serious harm to human welfare. The science is clear. But political leaders have not acted. Emissions grow to record levels every year. Science is no longer the bottleneck to action. The problem is social and political: politicians cannot act because there is not enough public support for the changes needed to secure our future. Changes in people's views and votes create the political support elected leaders require to act on the science. Changes in buying behaviour create incentives for businesses to transform their products and operations. The people cannot be ignored.

At the same time, action must remain grounded in and guided by the best available science. How can policymakers and the public learn about complex environmental issues, and do so in ways that promote action rather than causing denial or despair? How can learning occur when the size, complexity, time delays, and irreversible consequences of acting in many systems means there is no possibility of learning from experience? What is required both to generate reliable knowledge about complex problems, grounded in the best available scientific evidence, and to do so in ways that mobilize the grass roots support needed to implement high-leverage policies and motivate action to address highly politicized issues? Here I describe how systems thinking and interactive simulation modelling can help, focusing on the field of system dynamics (Forrester 1961, Randers 1980, Sterman 2000). I argue that interactive modelling is important, not only in generating knowledge, but also by catalysing the adoption of policies to address the pressing challenges we face. As an example I describe interactive simulations that are used around the world, from senior policymakers and business leaders to high school students, to build shared understanding of, and actions to address one of the most daunting and urgent issues we face: climate change.

13.1 POLICY RESISTANCE

Thoughtful leaders throughout society increasingly suspect that the policies we implement to address these and other difficult challenges have not only failed to solve the persistent problems we face, but are in fact causing them. All too often, well-intentioned programmes create unanticipated "side effects". The result is *policy resistance*, which is the tendency for well-intended interventions to be defeated by the system's broader responses to the intervention itself. Examples abound: Fossil fuels power our economy but harm human welfare through air pollution and climate change. The overuse of antibiotics spreads resistant pathogens. Radar, sonar, GPS and other technological innovations that boost catch per boat cause faster collapse of fish stocks, forcing catch down. Prosperity should improve human welfare, but the sedentary lifestyles and cheap calories that prosperity affords have led to epidemics of obesity and diabetes. Our best efforts to solve problems often make them worse (Sterman 2000, 2012). We try harder – and find ourselves even more stuck.

Policy resistance arises from a failure of systems thinking, from a narrow, reductionist worldview. We have been trained to view our situation as the result of forces outside ourselves, forces largely unpredictable and uncontrollable. Consider the "unanticipated events" and "side effects" so often invoked to explain policy failure. Political leaders blame recession on corporate fraud or terrorism. Managers blame bankruptcy on events outside their organizations and (they want us to believe) outside their control. But there are no side effects – just *effects*. Those we expected or that prove beneficial we call the main effects and claim credit. Those responses from the broader system that undercut our policies and come back to cause us harm we label side effects, hoping to excuse the failure of our intervention. "Side effects" are not a feature of reality but a sign that the boundaries of our mental models are too narrow, our time horizons too short.

13.2 COMPLEXITY, LEARNING FAILURES AND THE IMPLEMENTATION CHALLENGE

Generating reliable evidence through scientific method requires the ability to conduct controlled experiments, discriminate among rival hypotheses, and replicate results. But the more complex the issue, the more challenging these tasks become. Economic, social, medical, and other policies are embedded in intricate networks of physical, biological, ecological, technical, economic, social, political and other relationships. Experiments in complex human systems are often unethical or simply infeasible (we cannot release smallpox to test policies to thwart bioterrorists). Replication is difficult or impossible (we have only one climate and cannot compare a high greenhouse gas future to a low one). Decisions made in one part of a system quickly ripple out across geographic and disciplinary boundaries. Long time delays mean we never experience the full consequences of our actions. Follow-up studies must be carried out over decades or lifetimes, while changing conditions may render the results irrelevant. In summary, as systems become more complex and time frames grow longer, the more potential confounding factors arise, making it harder to find the causal needle in the haystack of spurious correlations. Complexity hinders the *generation* of evidence.

Learning often fails even when clear evidence is available. More than two and a half centuries passed from the first demonstration that citrus fruits prevent scurvy until citrus use was mandated in the British merchant marine, despite the importance of the problem and unambiguous evidence supplied by controlled experiments (Mosteller 1981). Some argue that today we are smarter and learn faster. Yet, for example, adoption of medical treatments varies widely across regions, socioeconomic strata, and nations, indicating overuse by some or underuse by others – despite access to the same evidence on risks and benefits (Fisher *et al.* 2003, Clancy and Cronin 2005). From airline kitchens to health care, similar firms in the same industry and even different floors of the same hospital exhibit persistent performance differences. These differences persist despite the presence of financial incentives, market forces, publications, benchmarking, training and imitation that should – in theory – lead to broad diffusion of best practices (Gibbons and Henderson 2013, Wennberg 2010). For example, total factor productivity varies by about a factor of two between the 10th and 90th percentile firms in the same industries in the US, and by a factor of more than five in China and India (Syverson 2011). Complexity hinders *learning* from evidence.

Many scientists respond to the complexity and learning problems by arguing that policy making should be left to the experts. But the Manhattan Project approach, in which experts secretly provide advice to inform decisions made without consulting the public, fails when success requires understanding and behavioural change throughout society. Policies to manage complex natural and technical systems should be based on the best available scientific knowledge. However, the beliefs of the public, not only those of experts, shape and constrain government policy. When science conflicts with "common sense" people are unlikely to favour or adopt policies consistent with science (Fischhoff 2007, 2009, Morgan et al. 2001, Slovic 2000, Bostrom et al. 1994, Read et al. 1994).

Abundant evidence shows that seat belts, motorcycle helmets, and childhood vaccinations save lives, yet legislation mandating their use took decades. Citizen groups campaign actively against many of these policies, and compliance remains spotty. The connection between actions and outcomes in these cases is far simpler than the connection between GHG emissions, climate and human welfare. People are often suspicious of experts and their evidence, believing – often with just cause – that those with power and authority routinely manipulate the policy process for ideological, political, or pecuniary purposes.

From motorcycle helmets to routine immunization, noncompliance and active resistance arise wherever people are unable to assess the reliability of evidence about complex issues on their own and are excluded from the policy process. Without effective risk communication, even the best science creates a knowledge vacuum that is then filled by error, disinformation and falsehood – some supplied inadvertently by people without knowledge of the science and some injected deliberately by ideologues and vested interests (Oreskes and Conway 2010). Complexity hinders the *implementation* of policies based on evidence.

DYNAMIC COMPLEXITY

Most people define complexity in terms of the number of components or possible states in a system. For instance, optimally scheduling an airline's flights and crews to provide good service at low cost is highly complex, but the complexity lies in finding the best solution out of an astronomical number of possibilities. Such problems have high levels of *combinatorial* complexity. However, most cases of policy resistance arise from *dynamic complexity* – the often counterintuitive behaviour of complex systems that arises from the interactions of the agents over time. Where the world is dynamic, evolving, and interconnected, we tend to make decisions using mental models that are static, narrow, and reductionist. Research shows that the elements of dynamic complexity people find most perplexing are feedback, time delays, and accumulations (stocks and flows).

FEEDBACK

Like organisms, the environment, economy and society are embedded in intricate networks of feedback processes, both self-reinforcing (positive) and self-correcting (negative) loops. However, studies show people have limited cognitive capacity and recognize only a few feedbacks. Thus, people usually think in short causal chains, tend to assume each effect has a single cause and often cease their search for explanations as soon as the first sufficient cause is found (Dörner 1996, Plous 1993).

For example, the US Corporate Average Fleet Efficiency (CAFE) regulations require the average efficiency of new vehicles to reach 54.5 miles per gallon (mpg) by 2025, a 237% improvement relative to the average US light duty fleet efficiency of approximately 23 mpg in 2012. Such a large improvement would, it is argued, cut US gasoline consumption and GHG emissions substantially, as suggested by Figure 13.1.

FIGURE 13.1: Open-loop mental model: Tightening vehicle efficiency standards should raise average fleet efficiency, lowering oil consumption and GHG emissions. No feedbacks or so-called "side effects" of higher efficiency are recognized. Arrows indicate causal influence; arrow polarity, e.g., x→⁺y indicates an increase in x raises y above what it would have been otherwise; x→⁻y indicates an increase in x lowers y below what it would have been otherwise (Sterman 2000, Ch. 5 provides formal definitions and examples).

Policy resistance arises because we do not understand the full range of feedbacks surrounding – and created by – our decisions. Higher vehicle efficiency is critical if we are to cut our greenhouse gas emissions. But the connection between efficiency standards, petroleum consumption and GHG emissions is not as simple as the open loop model in Figure 13.1 suggests. Figure 13.2 shows a few of the feedbacks affecting vehicle efficiency and petroleum consumption. As efficiency rises, petroleum demand will drop, causing oil prices to fall, leading people to buy larger, less efficient vehicles. Lower prices will also stimulate higher oil consumption in other sectors of the economy (aviation, shipping, heating and industries using oil as feedstocks). Lower oil prices will cut investment in both energy efficiency and in the renewable energy sources we need to cut petroleum use and GHG emissions. Less investment in renewables limits the scale economies and process improvements that lower their costs, further reducing their attractiveness and production in a vicious cycle.

As higher efficiency and lower fuel prices lower the cost of driving (per vehicle mile), people will carpool less, use less mass transit, and drive more, undercutting the benefits of the greater efficiency, a process known as the direct rebound effect. As mass transit revenues fall, transit systems will be forced to raise fares or cut service, driving still more people into cars and further cutting transit system revenue, another vicious cycle (Sterman 2000, Ch. 5). Lower oil prices will erode public support for strict efficiency standards, allowing large gas-guzzlers to stay on the market. The savings from efficiency and lower oil prices boost people's real incomes and may increase spending on other goods and services, increasing people's energy use and GHG emissions elsewhere in the economy, a process known as the indirect rebound effect (Herring and Sorrell 2009, Sorell et al. 2009). Indeed, spending our gasoline savings on, say, vacations and the flights to get there may even cause our total carbon footprint to rise. The impact of efficiency standards on oil consumption and GHG emissions is determined by a complex

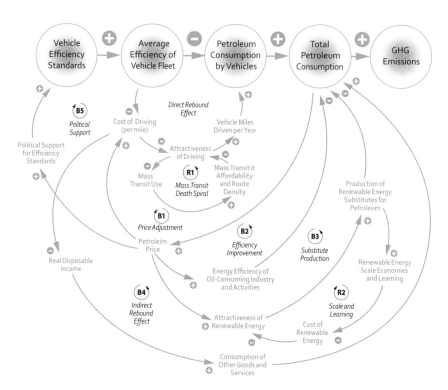

FIGURE 13.2:   A few of the feedbacks affecting efficiency, oil consumption and GHG emissions. Feedback loops are shown by loop identifiers and names. Balancing (negative) feedbacks are shown by "B"; reinforcing (positive) feedbacks are shown by "R". (see Sterman 2000 for formal definitions and examples). Delays are not shown. However, in contrast to such open-loop reasoning, the world reacts to our interventions. There is feedback: Our actions alter the environment, which in turn affects the decisions we make tomorrow. Our actions may trigger so-called side effects we didn't anticipate. Other agents, seeking to achieve *their* goals, act to restore the balance we have upset. Yesterday's solutions generate today's problems.

network of feedbacks, both balancing and self-reinforcing. Many of these feedbacks offset the intended benefits of efficiency standards.

The existence of these feedbacks does not mean that greater vehicle efficiency is unimportant in the quest for a renewable, carbon-free energy system. To the contrary, creating markets for alternative fuel vehicles that are sustainable both ecologically and economically is essential in limiting GHG emissions and the risks of climate change. But the idea that there is a simple technical fix for any environmental problem is both wrong and dangerous. Effective policies must account for the widest range of feedbacks surrounding the impact of "obvious" solutions, lest we be blindsided by unintended consequences that undermine their intended benefits and erode the political support needed for them to remain in force long enough to work.

TIME DELAYS

Time delays in feedback processes are common and particularly troublesome. Most obviously, delays slow the accumulation of evidence. More problematic, the short and long-run impacts of our policies are often different (smoking gives immediate pleasure while lung cancer develops over decades). Delays also create instability and fluctuations that confound our ability to learn. Driving a car, drinking alcohol, and building a new semiconductor plant all involve time delays between the decision to act (accelerating/braking, deciding to "have another", deciding to build a new factory) and its effects on the state of the system. As a result, decision-makers often continue pushing to correct apparent discrepancies between the desired and actual state of the system even after sufficient corrective actions have been taken to restore balance. The result is overshoot and oscillation: stop-and-go traffic, drunkenness, oil-price swings, shipping cycles, and high-tech boom and busts (Sterman 1989, Randers and Gölüke 2007, Gölüke this volume).

People routinely ignore or underestimate time delays (Sterman 1989, 2000). Underestimating time delays leads people to believe, wrongly, that it is prudent to "wait and see" whether a potential environmental risk will actually cause harm. Many citizens, including many who believe climate change poses serious risks, advocate a wait-and-see approach. Since there is uncertainty about the causes and consequences of climate change, they argue, we should defer potentially costly actions to address the risks. If climate change turns out to be more harmful than already seen, they reason, policies to mitigate it can be implemented at that time.

Wait-and-see policies often work well in simple systems, specifically those with short time lags between the detection of a problem, the implementation of corrective policies, and the impact of those actions. For instance, when boiling water for tea, one can wait until the kettle boils before removing it because there is essentially no delay between the boiling of the water and the whistle of the kettle, nor between hearing the whistle and switching it off. Similarly, wait-and-see policies can only be a prudent response to the risks of climate change if there are short delays in all the links of a long causal chain. That chain runs from the detection of adverse climate impacts to the implementation of mitigation policies to the resulting emissions reductions to changes in atmospheric GHG concentrations to global warming to changes in ice cover, sea level, weather patterns, agricultural productivity, habitat loss, extinction rates, and other impacts. Contrary to the logic of "wait and see" there are long delays in every link of the chain.

To make things even more problematic, the short and long-run impacts of policies are often different (Forrester 1969, Sterman 2000, Repenning and Sterman 2001). Such "Worse Before Better" and "Better Before Worse" behaviour is common: credit card debt boosts consumption today but forces austerity when the bills come due; restoring a depleted fishery requires cutting the catch today. The trade-off between short and long-run responses is particularly difficult in the context of climate change because the lags are exceptionally long. Long lags make standard frameworks for tradeoffs between short and long-run impacts such as discounting problematic. Even with low discount rates, potentially catastrophic events sufficiently far in the future, such as sea level rise from the loss of the Greenland or West Antarctic ice sheets, are given essentially no weight. Further, people commonly exhibit inconsistent time preferences (Frederick, Loewenstein and O'Donoghue 2002). For example, people often prefer two candy bars in 101 days over one candy bar in 100 days, but prefer one bar today over two tomorrow, a violation of standard assumptions of rational decision theory. The preference for immediate gratification, which appears to have a neurological basis (McClure et al. 2004), often leads people to avoid actions with long-term benefits they themselves judge to be desirable, over and above the usual effects of discounting.

STOCKS AND FLOWS

The process of accumulation – stocks and flows – is fundamental to understanding dynamics in general and the climate-economy system in particular. The stock of CO_2 in the atmosphere accumulates the flow of CO_2 emissions less the flow of CO_2 from the atmosphere into biomass and the ocean. The mass of the Greenland ice sheet accumulates snowfall less melting and calving. The stock of coal-fired power plants is increased by construction and reduced by decommissioning. And so on.

People should have good intuitive understanding of accumulation because stocks and flows are pervasive in everyday experience. Our bathtubs accumulate the inflow of water through the faucet less the outflow through the drain, our bank accounts accumulate deposits less withdrawals, and we all struggle to control our weight by managing the inflows and outflows of calories through diet and exercise. Yet research shows that people's intuitive understanding of stocks and flows is poor in two ways that cause error in assessing climate dynamics. First, people have difficulty relating the flows into and out of a stock to the level of the stock, even in simple, familiar contexts such as bank accounts and bathtubs. Second, narrow mental model boundaries mean people are often unaware of the networks of stocks and flows in a system.

Poor understanding of accumulation leads to serious errors in reasoning about climate change. Sterman and Booth Sweeney (2007) gave graduate students at MIT a description of the relationships among GHG emissions and atmospheric concentrations excerpted from the summary for policymakers in the IPCC's Third Assessment Report. Participants were then asked to sketch the emissions trajectory required to stabilize atmospheric CO_2 by 2100 at various concentrations. To draw attention to the stock-flow structure, participants were first directed to estimate future net removal of CO_2 from the atmosphere (net CO_2 taken up by the oceans and biomass), then draw the emissions path needed to stabilize atmospheric CO_2. The dynamics are easily understood without knowledge of calculus or climate science by applying a bathtub analogy. Figure 13.3 shows how the stock of CO_2 in the atmosphere rises when the inflow to the tub (emissions) exceeds the outflow (net removal). It is unchanging when inflow equals outflow, and falls when outflow exceeds inflow. Yet, 84% violated these elementary principles of accumulation. Most (63%) erroneously asserted that stabilizing emissions above net removal would

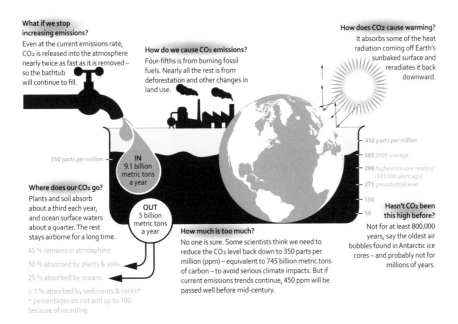

FIGURE 13.3: Portraying stocks and flows: The "Carbon Bathtub" (Source: *National Geographic*, December 2009; available at ngm.nationalgeographic.com/big-idea/05/carbon-bath).

stabilize atmospheric CO_2. That assertion is equivalent to arguing that a bath-tub continuously filled faster than it drains will never overflow. The false belief that stabilizing emissions would quickly stabilize the climate violates mass balance, one of the most basic laws of physics. And it also leads to complacency about the magnitude and urgency of emissions reductions required to mitigate climate change risk (Sterman 2008).

It might be argued that people understand the principles of accumulation but don't understand the carbon cycle or climate context. But the same errors arise in familiar settings such as bathtubs and bank accounts (Booth Sweeney and Sterman 2000, Sterman 2002, Cronin, Gonzalez and Sterman 2009). Moreover, training in science does not prevent these errors. Three-fifths of the participants had degrees in Science, Technology, Engineering or Mathematics (STEM); most others were trained in economics. These individuals are demographically similar to, and many will become, influential leaders in business and government, though with more STEM training than most. Merely providing more information on the carbon cycle will not alter the common but false belief that stabilizing emissions would quickly stabilize the climate.

13.3 LEARNING FOR OURSELVES
THROUGH INTERACTIVE SIMULATIONS

There is no learning without feedback. Feedback gives knowledge of the results of our actions. Scientists usually generate that feedback through controlled experimentation. Science progresses through an iterative process through which intuitions are challenged, hypotheses tested, insights generated, new experiments run. When experiments are impossible, as in the climate-economy system, scientists rely on models and simulations, which enable that iterative process to take place through controlled experimentation in virtual worlds (Sterman 1994, Edwards 2010). Learning arises in the process of interacting with the models, hypothesizing how the system might respond to policies, being surprised, forming new hypotheses, testing these with new simulations and data from the real world.

Paradoxically, however, scientists, having deepened their understanding through an iterative, interactive learning process, often turn around and tell the results to policymakers and the public through reports and presentations. Then they expect those leaders and the public to change their beliefs and behaviour, and even express surprise when these groups – excluded from the learning process, unable to assess the evidence on their own and presented with claims that conflict with deeply held beliefs – resist the message and challenge the authority of the experts.

When experimentation is impossible, simulation becomes the main – perhaps the only – way we can discover *for ourselves* how complex systems work and build broad, shared understanding of the long-term impact of different policies and thus integrate science into decision-making.

Let's look more closely at the case of climate change. In 1992 the nations of the world created the United Nations Framework Convention on Climate Change (UNFCCC) to negotiate binding agreements to address the risks of climate change. Nearly every nation on Earth committed to limiting global greenhouse gas (GHG) emissions to prevent "dangerous anthropogenic interference in the climate system".[1] Doing so is generally accepted to mean limiting the increase in mean global surface temperature to 2°C above preindustrial levels. The IPCC's Fifth Assessment Report (2014) concludes:

> *Continued emission of greenhouse gases will cause further warming and long-lasting changes in all components of the climate system, increasing the likelihood of severe, pervasive and irreversible impacts for people and ecosystems. Limiting climate change would require substantial and sustained reductions in greenhouse gas emissions which, together with adaptation, can limit climate change risks.[2]*

High hopes were dashed at the 2009 Copenhagen climate conference when face-to-face negotiations among heads of state collapsed. Instead, nations were encouraged to make voluntary pledges to reduce their emissions. These proved to be grossly inadequate: global emissions have risen to record levels since the great recession of 2008. Promising developments such as the 2014 US–China agreement and Lima Accord offer the best hope to date for emissions reductions, but still call on every nation to set its own target and timeline for emissions reductions. As of this writing these remain far too small to have a chance of achieving the 2° target.

The failure of global negotiations to date can be traced to the gap between the strong scientific consensus on the risks of climate change and widespread confusion, complacency and denial among policymakers, the media and the public (Sterman 2011). Even if policymakers fully understood the risks and dynamics of climate change – and many do not – in democracies, at least, the ratification of international agreements and passage of legislation to limit GHG emissions requires grassroots political support.

---

1   http://unfccc.int/essential_background/convention/background/items/1349.php

2   IPCC *Synthesis Report, SYR-18,* http://www.ipcc.ch/report/ar5/syr/

Historically, information about climate dynamics and risks comes to policymakers, negotiators and the public in the form of reports. These are based on the results of advanced general circulation models, such as those used by the IPCC. Such models are essential in developing reliable scientific knowledge of climate change and its impacts. However, these models are opaque and expensive, and neither available to nor understandable by non-specialists. The cycle time for creating and running scenarios is too long to allow real-time interaction with the models. Consequently, policymakers, educators, business and civic leaders, the media and the general public rely on their intuition to assess the likely impacts of emissions reduction proposals. However, as shown above, intuition, even among experts, is highly unreliable when applied to judging how proposals affect likely future GHG concentrations, temperatures, sea level, and other impacts.

Poor understanding of complex systems not only afflicts the public, but the negotiators themselves. In 2008, Christiana Figueres, then lead negotiator for Costa Rica, and named executive secretary of the UNFCCC in 2010, commented

> *Currently, in the UNFCCC negotiation process, the concrete environmental consequences of the various positions are not clear to all of us … There is a dangerous void of understanding of the short and long term impacts of the espoused … unwillingness to act on behalf of the Parties. (personal communication, Sept. 2008)*

The C-ROADS (Climate Rapid Overview And Decision Support) model is an interactive tool designed to address these challenges. C-ROADS helps build shared understanding of climate dynamics. It is solidly grounded in the best available science, rigorously nonpartisan, and understandable by and useful to non-specialists, from policymakers to the public.

In brief, C-ROADS:

- is based on the best available peer-reviewed science and calibrated to state-of-the-art climate models;
- tracks GHGs including CO_2, CH_4, N_2O, SF_6, halocarbons, aerosols and black carbon through the end of the century;
- distinguishes emissions from fossil fuels and from land use and forestry policies;
- allows users to select different business as usual scenarios, e.g., the IPCC Representative Concentration Pathways (RCPs), or to define their own;
- enables users to capture any emissions reduction scenario for each nation portrayed;

◗ reports the resulting GHG concentrations, global mean temperature change, sea level rise, ocean pH, per capita emissions, and cumulative emissions;

◗ allows users to assess the impact of uncertainty in key climate processes;

◗ is easy to use, running on a laptop computer in about one second so users immediately see the impact of the scenarios they test;

◗ provides an independent, neutral process to ensure that different assumptions and scenarios can be made available to all parties;

◗ is freely available at climateinteractive.org.

A TOOL FOR SCIENCE BASED ACTIVISTS:

C-ROADS is a continuous time model with an explicit carbon cycle, atmospheric stocks of other GHGs, radiative forcing, global mean surface temperature, sea level rise and surface ocean pH. Figure 13.4 shows the overall model architecture, and Figure 13.5 shows the stocks and flows capturing the carbon cycle. C-ROADS explicitly models CO_2 and other GHGs, including methane (CH_4), nitrous oxide (N_2O), SF_6 and other fluorinated gases (PFCs and HFCs), each with its own emissions fluxes, atmospheric stock and lifetime. The model structure and behaviour are described in detail in Sterman et al. (2013).[3]

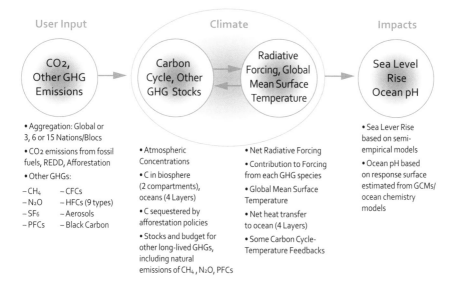

FIGURE 13.4:   C-ROADS Overview. User-specified scenarios for GHG emissions affect atmospheric concentrations and the climate, which in turn drive impacts including sea level and ocean pH. The model includes climate-carbon cycle feedbacks.

---

3   The complete documentation for C-ROADS is available at climateinteractive.org.

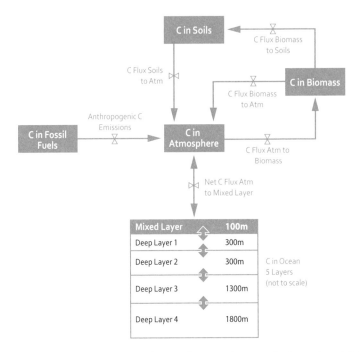

FIGURE 13.5: C-ROADS carbon cycle. CH$_4$ fluxes and atmospheric stock and C fluxes and stocks due to deforestation/afforestation are represented explicitly but are aggregated in this simplified view.

C-ROADS includes a variety of climate-carbon cycle feedbacks, including feedbacks from global mean temperature to net primary production and ocean CO$_2$ uptake. C-ROADS also includes positive feedbacks involving methane release from melting permafrost, but sets the base-case gains of these feedbacks by default to zero because they are, at present, poorly constrained by data. Consequently, C-ROADS is likely to underestimate future warming and sea level rise. Users can test any values they wish for these feedbacks. We revise the model as knowledge of climate-carbon cycle feedbacks improves.

C-ROADS simulations begin in 1850. The model is driven by historic CO$_2$ and GHG emissions and includes the impact of volcanoes, variations in insolation and other forcings. Figure 13.6 compares C-ROADS to data. The model tracks the data well (Sterman et al. 2013). The full model documentation compares C-ROADS to history for other GHGs and radiative forcing, and to other projections and models.

The user interface enables rapid experimentation with different policies and parameters. On the main screen users can access instructions, a video tutorial, interactive model structure diagrams and documentation, then select

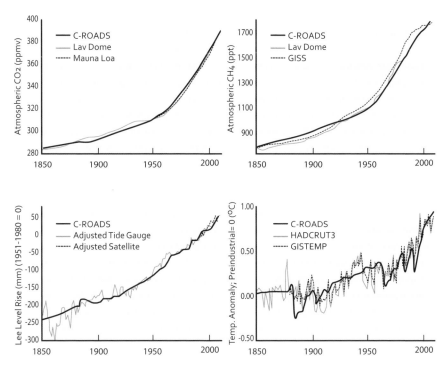

FIGURE 13.6:   C-ROADS fit to historical data. Clockwise from top left: Atmospheric CO_2, CH_4, temperature anomaly, sea level.

the level of regional aggregation for emissions, including global totals, or 3, 6, or 15 different nations and regional blocs (Table 13.1). Users interested in examining the impact of emissions from nations not explicitly represented can do so by developing a spreadsheet specifying the emissions projections for these nations; C-ROADS can read such files directly. Users then select a business as usual scenario, choosing those of the IPCC or Energy Modeling Forum, or specifying their own. Users can also load prior simulations, carry out Monte-Carlo sensitivity analysis to assess uncertainty and analyse the contribution of any nation's proposals to global outcomes.

Next, users define scenarios for anthropogenic CO_2 emissions from fossil fuels and land use and emissions of other GHGs through 2100 for individual countries and regional blocs (Figure 13.7). Users enter projected emissions for each nation or bloc in one of three modes: numerically, graphically, or from an Excel spreadsheet. Users can specify future emissions relative to a user-selected base year (e.g., "emissions in 2020 will be 17% below the 2005 value"); relative to the business as usual scenario (e.g., "emissions in 2020

TABLE 13.1:   In addition to the global level, C-ROADS users may choose 3, 6 or 15 nation/region levels of aggregation.

3 Regions[a]	6 Regions	15 Regions
Developed	China	Australia
All developed nations	European Union	Brazil
Developing A	India	Canada
Rapidly developing nations	United States	China
(Brazil, China, India, Indonesia,	Other Developed Nations	European Union
Mexico, South Africa and other	Australia, Canada, Japan, New	India
large developing Asian nations)	Zealand, Russia/FSU/ Eastern	Indonesia
Developing B	Europe, South Korea	Japan
Rest of world: least developed	Other Developing Nations	Mexico
nations in Africa, Asia, Latin	Brazil, Indonesia, Mexico, South	Russia
America, Middle East, Oceania	Africa; Other Africa, Asia, Latin	South Africa
	America, Middle East, Oceania	South Korea
		United States
		Developed non MEF[b] nations
		Other Eastern Europe & FSU,
		New Zealand
		Developing non MEF nations
		Other Africa, Asia, Latin Amer-
		ica, Middle East, Oceania

[a] The three region level of aggregation is available in C-Learn, the online version of C-ROADS.

[b] Major Economies Forum on Energy and Climate; www.majoreconomiesforum.org.

will be 30% below the scenario value for that year"); relative to the carbon intensity of the economy for each nation or bloc (e.g., "emissions in 2020 will reflect a 45% reduction in carbon intensity relative to 2005"); relative to per capita emissions for that nation or bloc (e.g., "emissions in 2050 will reflect 10% growth in emissions per capita over the 2005 level for that nation or bloc"); and more options (detailed in the documentation). Input modes, target years and emissions in each target year can differ for each nation and bloc.

Model output updates immediately. Users can select graphs and tables to display, by nation/bloc or globally, population and GDP, emissions of CO_2 and other GHGs, emissions per capita, the emissions intensity of the economy, CO_2 and CO_2e concentrations, CO_2 removal from the atmosphere, global mean surface temperature, sea level rise, ocean pH, and other indicators.

C-ROADS also offers interactive sensitivity analysis. Users can alter the values of key parameters, individually or in combination, and get immediate results.

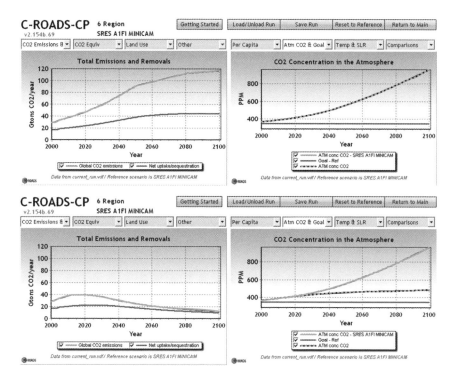

FIGURE 13.7: C-ROADS interface, showing emissions (left graph; 6 region level of aggre-gation) and global average surface temperature (right graph, °C above preindustrial levels). Tabs provide users with different ways to enter emissions pathways and proposals for each nation and bloc, pathways for GHGs other than CO_2, the ability to select different scenarios for population and economic output per capita, sensitivity analysis, and a variety of other graphs, tables, and options so they can carry out the experiments and use the assumptions they want to test.

APPLICATIONS

Negotiators, policymakers, scientists, business leaders, and educators are among the many who use C-ROADS. Senior members of the US government includ-ing legislators and members of the executive branch have used C-ROADS. The US Department of State Office of the Special Envoy for Climate Change has developed an in-house capability to use C-ROADS and deploy it in the UNF-CCC and other bilateral and multilateral negotiations. US Secretary of State, John Kerry, has used the model, and commented at a C-ROADS presentation in 2009 (when he was Chair of the US Senate Foreign Relations Committee)

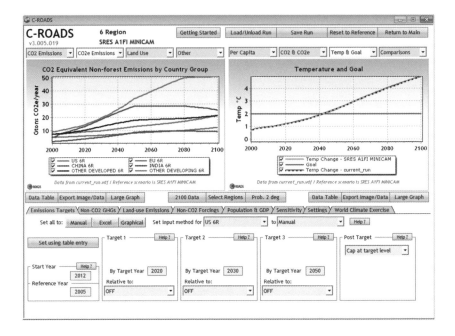

FIGURE 13.8:   Carbon mass balance or "bathtub dynamics" illustrated by C-ROADS. Top: The graph on the left shows the inflow to the stock of atmospheric CO_2 (global CO_2 emissions; red line) and the outflow of CO_2 from the stock in the atmosphere (net CO_2 removal as it is taken up by biomass and dissolves in the ocean; green line). The inflow always exceeds the outflow, so the level of CO_2 in the atmosphere rises continuously. The gap between inflow and outflow increases over time, so concentrations rise at an increasing rate, reaching 965 ppm by 2100. To stabilize atmospheric CO_2 concentrations, emissions must fall to net removal. Bottom: A scenario in which global emissions peak around 2020 and fall to roughly a third of the 2005 flux by 2100, by which time emissions and net removal are nearly in balance, so that CO_2 concentrations nearly stabilize (in this scenario, at about 485 ppm). Note that net CO_2 removal from the atmosphere falls in the stabilization scenario: lower atmospheric CO_2 concentrations reduce net uptake of CO_2 by biomass and the oceans.

More chilling is the computer modelling [the C-ROADS team] did against the current plans of every single country that is planning to do anything, and it's not that big a group … They took all of these current projections and ran the computer models against what is currently happening in the science. And in every single case, it showed that we are not just marginally above a catastrophic tipping point level. We are hugely, significantly above it.

Dr. Jonathan Pershing, the former Deputy Special Envoy at the State Department, now Deputy Assistant Secretary of Energy, commented

> *The results [of C-ROADS] have been very helpful to our team here at the U.S. State Department … The simulator's quick and accurate calculation of atmospheric carbon dioxide levels and temperatures has been a great asset to us. … I have made use of the results in both internal discussions, and in the international negotiations … (personal communication)*

Former staff member Dr. Benjamin Zaitchik elaborates

> *… [P]olicy makers and negotiators need to have a reasonable sense of what a particular action will mean for global climate, when considered in the context of other actions and policies around the world. Previously, we would make these calculations offline. We'd download emissions projections from a reliable modelling source, input them to an excel spreadsheet to adjust for various policy options, and then enter each proposed global emissions path into a model like MAGICC to estimate the climate response. This method … was time consuming and opaque: in the end we had a set of static graphs that we could bring into a meeting, but we couldn't make quick adjustments on the fly. With C-ROADS, we can adjust policy assumptions in real-time, through an intuitive interface. This makes it much easier to assess the environmental integrity of various proposed emissions targets and to discuss how complementary emissions targets might achieve a climate goal … (personal communication)*

Former State Department staffer, Eric Maltzer, commented, "You have been the backbone of our analytic work here" (personal communication, 2013).

C-ROADS is also used in China, where it has been disaggregated to include drivers of CO_2 emissions at the provincial level using assumptions about total energy use and fuel mix, by the United Nations Environment Program (e.g., UNEP 2010, 2011), and by climate policy activists, including 350.org's founder, Bill McKibben:[4]

> *…the only people who really understand what's going on may be a small crew of folks from a group of computer jockeys called Climate Interactive. Their software speaks numbers, not spin – and in the end it's the numbers that count.*

---

4   http://www.theguardian.com/environment/cif-green/2009/dec/15/bill-mckibben

EDUCATIONAL USE

A free online version, C-Learn, is widely used in classrooms. C-ROADS and C-Learn are also used in an interactive role-play simulation of the global climate negotiations entitled *World Climate* (Sterman et al. 2014). Instructions and all materials needed to run World Climate are freely available at climateinteractive.org.

Participants playing the roles of major nations negotiate proposals to reduce emissions, using C-ROADS to provide immediate feedback on the impacts of their proposals. Participants learn about the dynamics of the climate and impacts of proposed policies in a way that is consistent with the best available peer-reviewed science but that does not prescribe what should be done.

For example, participants commonly believe stabilizing atmospheric CO_2 concentrations and the climate requires only that emissions be stabilized. As discussed above, this erroneous belief arises from people's poor understanding of stocks and flows. C-ROADS enables people to discover the dynamics of accumulation for themselves. Initial proposals often stabilize emissions well above current levels. When they simulate these proposals, however, they find that CO_2 concentrations steadily grow, because the "Carbon Bathtub" (Figure 13.3) is still filled faster than it drains. Through experimentation, they discover that stabilizing atmospheric CO_2 concentrations requires substantial emissions cuts, and soon (Figure 13.8).

World Climate has been used successfully with groups including students, business executives and political leaders. Grass-roots civil society organizations such as the youth-led "COPinMyCity"[5] use C-ROADS and World Climate to educate and inspire. The COPinMyCity programme combines *World Climate* with "mobilization, awareness raising and debriefing" – what they call "Simul-action" – to help participants connect the science to action.[6]

Evaluations show *World Climate* improves participant knowledge of climate science and policy options (Sterman et al. 2014). Even more important, the experience can generate hope and catalyse action, as these participant comments, collected in sessions run by Prof. Juliette Rooney-Varga, Univ. of Massachusetts, Lowell, illustrate:

> *This exercise makes me think I will have to tackle climate change more seriously.*

> *I feel surprisingly excited. All of these problems can be solved … We have a chance to build a new world.*

---

5   http://copinmycity.weebly.com
6   http://copinmycity.weebly.com/simulating-the-nego.html

I will do what I can to consume less and motivate the people close to me to do the same.

I would like to reduce my own CO_2 emission[s] and lead a programme for this in my future career.

LIMITATIONS AND EXTENSIONS

C-ROADS enables decision-makers, educators, the media, and the public to quickly assess important climate impacts of particular national, regional or global emissions scenarios and to learn about the dynamics of the climate.

As with any model, C-ROADS is not appropriate for all purposes. To be able to run in about a second on standard laptops, the carbon cycle and climate sectors are globally aggregated. Thus C-ROADS cannot be used to assess climate impacts at regional or smaller scales.

C-ROADS takes future population, economic growth, and GHG emissions as scenario inputs specified by the user and currently omits the costs of policy options and climate change damage. Many users, particularly those involved in negotiations, value the ability to specify pledges and proposals exogenously, i.e. as external inputs to the model. But GHG emissions result from complex interactions of energy demand, production, prices, technology, learning and scale economies, regulations and government policies.

To address these issues, the Climate Interactive developed the En-ROADS model. En-ROADS endogenously generates energy use, fuel mix, and GHG emissions. The model represents energy producing capital stocks such as oil wells and power plants, and energy consuming capital stocks such as vehicles and buildings. The model portrays the extraction, processing, and consumption of oil, gas, coal, nuclear power, and renewables (e.g., hydropower, wind, solar, biofuels). The capacity to produce each energy source, and the number and efficiency of vehicles and buildings that consume energy, depends on investment, which in turn depends on the profitability of each fuel type.

En-ROADS includes construction and planning delays for the development of new energy sources and the possibility of retrofits and early retirement for existing capital stocks. The costs, prices and profitability of each energy source are endogenous. Costs vary with resource depletion and supply bottlenecks that can raise costs, and R&D, learning curves, and other feedbacks that can lower costs. Users can test a wide range of policies including carbon prices, regulatory constraints and subsidies for specific technologies. Users can also vary important assumptions determining energy resource availability, technical breakthroughs, cost reductions, construction times and lifetimes for new plants, the potential for efficiency and retrofits, etc. The resulting emissions are then the input to the same carbon cycle and climate structures used in C-ROADS, so users immediately see the impact of policies such as carbon prices, efficiency standards, subsidies for renewables, etc., on emissions, GHG concentrations, global average temperatures, sea level rise and other climate impacts. Like C-ROADS, En-ROADS simulates in seconds on an ordinary laptop.

CONCLUSION

Policies to address pressing challenges often fail or worsen the problems they are intended to solve. Evidence-based learning should prevent such policy resistance, but learning in complex systems is often weak and slow. Complexity hinders our ability to discover the delayed and distal impacts of interventions, generating unintended so-called side effects. Yet learning often fails even when strong evidence is available: common mental models lead to erroneous but self-confirming inferences, allowing harmful beliefs and behaviours to persist and undermining implementation of beneficial policies. When evidence cannot be generated through experiments in the real world, virtual worlds and simulation become the only rigorous way to test hypotheses and evaluate the potential effects of policies.

Most important, when experimentation in real systems is infeasible, simulation is often the only way we can discover for ourselves how complex systems work. Without the rigorous testing enabled by simulation, it becomes all too easy for policy to be driven by unconscious bias, superstition or ideology. The alternative is rote learning based on the authority of an expert, a method that dulls creativity and stunts the development of the skills needed to catalyse effective change in complex systems.

Interactive simulations such as the models developed by Climate Interactive, and by MIT Sloan[7] enable people – from senior political and business leaders to students – to develop their systems thinking capabilities and understand the science and politics of complex issues. Through such evidence-based, interactive simulations, we can then use these capabilities to design better policies and build the collective commitment needed to implement them.

---

7   https://mitsloan.mit.edu/LearningEdge/simulations

BIBLIOGRAPHY

Booth Sweeney, L. and Sterman, J. (2000). Bathtub Dynamics: Initial Results of a Systems Thinking Inventory. *System Dynamics Review, 16*(4), 249–294.

Bostrom, A., Morgan, M.G., Fischhoff, B. and Read, D. (1994). What Do People Know About Global Climate Change? Part 1: Mental models, *Risk Analysis, 14*(6), 959–970.

Clancy C.M. and Cronin, K. (2005). Evidence-Based Decision Making: Global Evidence, Local Decisions, *Health Affairs. 24*(1), 151–162.

Cronin, M., Gonzalez, C. and Sterman, J. (2009). Why Don't Well-Educated Adults Understand Accumulation? A Challenge to Researchers, Educators, and Citizens. *Organizational Behavior and Human Decision Processes, 108*(1), 116–130.

Dörner, D. (1996). *The Logic of Failure*. New York: Metropolitan Books/Henry Holt.

Edwards, P. (2010). *A Vast Machine*. Cambridge: MIT Press.

Fisher, E.S., Wennberg, D.E., Stukel, T.A., Gottlieb, D.J., Lucas, F.L., Pinder, E.L. (2003). The implications of regional variations in Medicare spending. Part 1: The content, quality, and accessibility of care. *Annals of Internal Medicine, 138*(4), 273–287.

Fischhoff, B. (2009). Risk Perception and Communication. In R. Detels, R. Beaglehole, M. Lansang, and M. Gulliford (Eds.), *Oxford Textbook of Public Health*, 5th Ed., pp. 940–952. Oxford: Oxford University Press.

Fischhoff, B. (2007). Non-Persuasive Communication about Matters of Greatest Urgency: Climate Change. *Environmental Science & Technology*, *41*, 7204–7208.

Forrester, J.W. (1961). *Industrial Dynamics*. Cambridge, MA: MIT Press.

Forrester, J.W. (1969). *Urban Dynamics*. Cambridge MA: MIT Press.

Frederick, S., Loewenstein, G. and O'Donoghue, T. (2002). Time Discounting and Time Preference: A Critical Review. *Journal of Economic Literature, 40*(2), 351–401.

Gibbons, R. and Henderson, R. (2013). What do managers do? In Gibbons, R. and Roberts, J. (Eds.). *Handbook of Organizational Economics,* pp. 680–731. Princeton: Princeton University Press.

Herring, H. and Sorrell, S. (2009). *Energy efficiency and sustainable consumption: the rebound effect*. London: Palgrave Macmillan.

McClure, S., Laibson, D., Loewenstein, G. and Cohen, J. (2004). Separate Neural Systems Value Immediate and Delayed Monetary Rewards. *Science, 306*(15 October), 503–507.

Meadows, D., Randers, J., Meadows, D. (2004). The limits to growth: the thirty year update. White River Junction, VT : Chelsea Green.

Morgan, G., Fischhoff, B., Bostrom, A. and Atman, C. (2001). *Risk Communication: A Mental Models Approach.* Cambridge, UK: Cambridge University Press.

Mosteller F. (1981). Innovation and evaluation, *Science, 211*, 881–886.

Oreskes, N. and Conway, E. (2010). *Merchants of Doubt.* Bloomsbury Press.

Plous, S. (1993). *The Psychology of Judgment and Decision Making.* New York: McGraw Hill.

Randers, J. (ed) (1980). *Elements of the System Dynamics Method. Cambridge.* MA: MIT Press.

Randers, J. (2012). *2052: A Global Forecast for the Next 40 Years.* White River Junction, VT: Chelsea Green.

Randers, J. and Göluke, U. (2007). Forecasting turning points in shipping freight rates: lessons from 30 years of practical effort. *System Dynamics Review, 23*(2/3), 253–284.

Read, D., Bostrom, A., Morgan, M.G., Fischhoff, B. and Smuts, D. 81994). What Do People Know About Global Climate Change? Part 2: Survey studies of educated laypeople, *Risk Analysis, 14*(6), 971–982.

Repenning, N. and Sterman, J. (2001). Nobody Ever Gets Credit for Fixing Problems that Never Happened: Creating and Sustaining Process Improvement. *California Management Review, 43*(4), 64–88.

Rockström, J. et al. (2009). A safe operating space for humanity. *Nature, 461*, 472–475.

Running, S. (2012). A Measurable Planetary Boundary for the Biosphere. *Science, 337*, 1458–1459.

Slovic, P. (ed.) (2000). *The Perception of Risk.* London: Earthscan.

Sorrell, S., Dimitropoulos, J. and Sommerville, M. (2009). Empirical estimates of the direct rebound effect: A review. *Energy Policy, 37*, 1356–1371.

Sterman, J. (1989). Modeling Managerial Behavior: Misperceptions of Feedback in a Dynamic Decision Making Experiment. *Management Science, 35*(3), 321–339.

Sterman, J. (1994). Learning In and About Complex Systems. *System Dynamics Review, 10*(2-3), 291–330.

Sterman, J. (2000). *Business Dynamics: Systems Thinking and Modeling for a Complex World.* Boston: Irwin/McGraw-Hill.

Sterman, J. (2002). All Models are Wrong: Reflections on Becoming a Systems Scientist. *System Dynamics Review, 18*(4), 501–531.

Sterman, J. (2008). Risk Communication on Climate: Mental Models and Mass Balance. *Science, 322*, 532–533.

Sterman, J. (2011). Communicating Climate Change Risks in a Skeptical World. *Climatic Change*, *108*, 811–826.

Sterman, J. (2012). Sustaining Sustainability: Creating a Systems Science in a Fragmented Academy and Polarized World. In M. Weinstein and R.E. Turner (Eds.), *Sustainability Science: The Emerging Paradigm and the Urban Environment*, Springer: 21–58.

Sterman, J. and L. Booth Sweeney (2007). Understanding Public Complacency About Climate Change: Adults' Mental Models of Climate Change Violate Conservation of Matter. *Climatic Change, 80*(3-4), 213–238.

Sterman, J., Fiddaman, T., Franck, T., Jones, A., McCauley, S., Rice, P., Sawin, E., Siegel, L. (2013). Management Flight Simulators to Support Climate Negotiations. *Environmental Modelling and Software*. *44*, 122–135.

Sterman, J. et al. (2014). *World Climate: A Role-Play Simulation of Global Climate Negotiations. Simulation and Gaming*. DOI: 10.1177/1046878113514935.

Syverson, C. (2011). What Determines Productivity? *Journal of Economic Literature, 49*, 326–365.

UNEP (2011), *Bridging the Emissions Gap. United Nations Environment Programme (UNEP)*. Available at www.unep.org/publications/ebooks/bridgingemissionsgap.

Wackernagel, M., Schulz, N., Deumling D., et al. (2002). Tracking the ecological overshoot of the human economy. *PNAS, 99*, 9266–9271.

Wennberg, J. (2010). *Tracking Medicine: A Researcher's Quest to Understand Health Care*. New York: Oxford University Press.

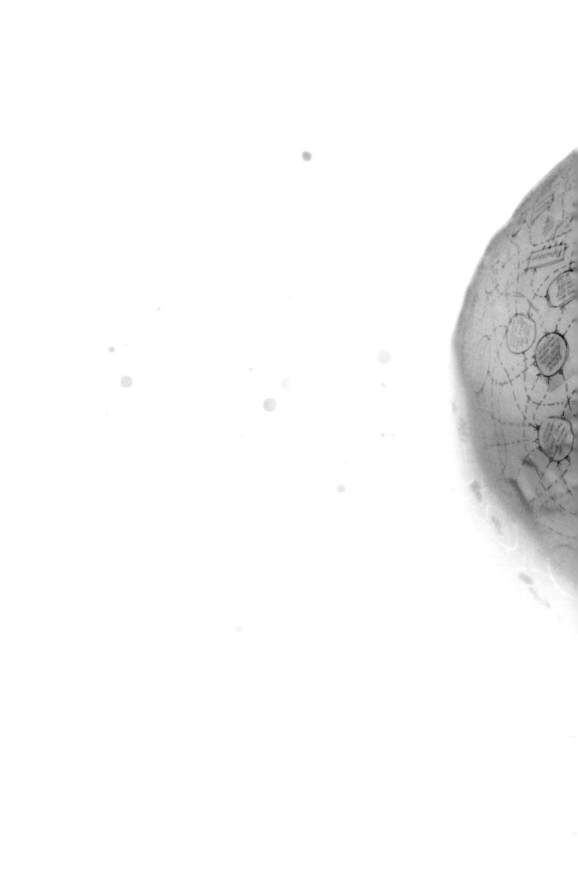

CHAPTER 14

System Dynamics of Shipping Cycles: How Does it Create Value?

By Ulrich Golüke

INTRODUCTION

Over many years, more than three decades in fact, Jorgen Randers and I worked with Norwegian shipping investors, advising them on when to buy or sell bulk vessels. We developed a system dynamics model that managed to shift the focus of the investors from external transport-demand forecasts to processes of decision-making internal to their mental models. We also found a mode of presenting that allowed the investors to rapidly internalize our model output, thus improving their decisions. This model demonstrates the usefulness of appropriate models for influencing decision-making from within large, uncertain socio-economic systems.

14.1 ELEMENTS OF SHIPPING

The transportation of goods by ship is probably as old as human history. The high seas are wide and open, the number of seafaring nations large and the return on investment in vessels significant and very volatile. Many fortunes can and have been made – and lost. Thus, an industry emerged that is highly fragmented to this day. As BIMCO, the Baltic and International Maritime Council, wrote: "Ownership of vessels is a concept hedged around with complexities."

In this paper I discuss the creation of value in bulk transportation through the buying and selling of "bulkers." The shipping industry calls this *wet* bulk – crude oil and oil products – and *dry* bulk – coal, iron ore, aluminum, grain and other dry bulk. Of the approximately 80 000 vessels above 1000 gross tonnage capacity about 20 000 are bulkers. The rest are general cargo ships, container vessels, ro-ro cargo ships, gas carriers, passenger ships, offshore vessels, service ships and tugs. Even though the *number* of bulkers is only 25% of all ships, by *tonnage* they represent about 60% of the total.

This fragmented industry serves countless customers, plies hundreds of routes and is owned by many players. The cost of transport is vanishingly small in comparison to the value of goods being transported, so the impression of an industry at the mercy of demand that is established elsewhere is very widespread amongst all the industry participants: investors, charterers, regulators, seamen and women.

As a result, much effort is spent on forecasting total demand and its regional and temporal distribution. Because the reward for getting it right is truly tremendous. Freight rates being paid to owners vary by a factor of five or more. At the very top of the market one can earn enough money with a few trips to pay for an entire new or secondhand vessel. The earnings throughout the rest of the vessels lifetime are, as they say, pure gravy.

On the other hand, when the rates are low, it is easy to lose a fortune.

14.2 FROM AN EXTERNAL
PERSPECTIVE TO AN INTERNAL ONE

The tendency to seek external explanations, i.e. demand which from the industry's perspective is uncontrollable, for freight rate movements rather than internal ones, i.e. supply over which the industry does have control, is exacerbated by the opacity of fundamental market data.

The industry can take two decisions: 1) how much capacity to have available – through *ordering* and *scrapping* of tonnage – and 2) how to *utilize* the capacity it has available at any one time. Of the resulting three options, how much to order, how much to scrap and how intensely to utilize capacity, only scrapping is known in a timely and accurate manner. This is in spite the fact that there are a finite, countable and visible number of vessels plying for trade. Or is it *because* of this fact? Quite possibly, because each vessel investment decision and even each trade is an individually negotiated agreement. Attempts at standardization, and thus comparability, exist but have not gotten very far – except for container trades.

Thus we are left with a situation where capacity utilization is strongly influenced by such soft, and very difficult to measure, factors like slow steaming, route planning, waiting for cargo, maintenance scheduling, and days-in-port. Together, these influences can adjust capacity utilization by a factor of two or more, and do so relatively fast.

Even the ordering of vessels is less clear than one would think. Again, despite the fact that the number of investors and the number of shipyards is finite, it is quite difficult to get an accurate picture of how much tonnage will enter the market at what time. This is not only due to the variable time it takes to build a ship, the order backlog and the efficiencies of shipyards, but also due to cancellations of orders and the stretching of delivery times at the request of the buyer. As a result, only two thirds of the tonnage reported as placed on order appears on average as real ships entering the market and thus increasing capacity.

Finally, one should not underestimate the psychological advantage, not only in shipping, in seeking and believing external reasons for one's own success and especially misfortune. We all prefer to put the blame elsewhere – it avoids having to take too close a look at oneself.

System dynamicists, on the other hand, look for endogenous explanations by nature (examples from shipping are Taylor 1976 and 1982, and Bakken 1993). Not in order to be able to better castigate oneself, but to be able to better identify leverage points in the system, so that with relatively little effort much can be changed – the result is this causal loop diagram:

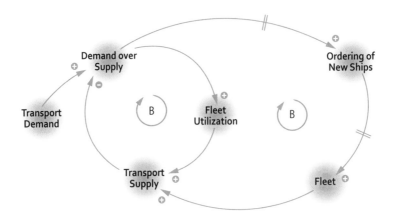

FIGURE 14.1

CAUSAL LOOP DIAGRAM

Causal loops in system dynamics are more than pretty drawings. They are concise graphical representations of a hypothesis about how a system works and behaves. This allows for more exact intervention in a system. If the system is an economic one, like shipping, this knowledge creates considerable value.

The variable "Ordering of New Ships" actually stands for a decision making process that tries to bring existing vessel tonnage in line with demanded, or needed, vessel tonnage in the future. To do so, the investor can order new tonnage or not, and if he or she is at the same time the owner of the vessel, he can in addition decide to scrap (old) tonnage. Scrapping, however, is a relatively orderly process determined overwhelmingly by the age of the vessel, and very often delegated to the operations department. We have never heard of a board meeting where scrapping decisions were discussed, in contrast to investment decisions, which often dominate the agenda.

One reads such diagrams in the following way:

▶ The variable at the tail of an arrow *causes* a change in the variable at the head. A "plus" sign at the head of the arrow signifies a change in the same direction; a "minus" sign signifies change in the opposite direction. For example, if "demand over supply" *increases*, then "fleet utilization" also *increases*. And in a second example, if "transport supply" *increases*, then "demand over supply" *decreases*.

▶ A double line across an arrow signifies a significant delay.

▶ The variables are arranged in loops, i.e. there are no dependent and independent variables, all are both.

▶   Loops are either "balancing" or "self-reinforcing." As an example, the inner loop above is balancing, because if, say, "demand over supply" *increases*, then "fleet utilization" *increases*; if "fleet utilization" *increases,* then "transport supply" *increases*; and if "transport supply" *increases,* "demand over supply" *decreases*, thus closing the loop. An initial increase in a variable causes a chain of changes that eventually causes a decrease in the originating variable, resulting in a balanced behavior over time. One can start with any variable in the loop, and one can assume an initial increase or decrease, the loop always returns to balance – just try for yourself.

BALANCING LOOPS

All systems that survive over time are dominated by balancing loops. If they were not, they would sooner rather than later collapse and die. Just recall the fable of the rice and the chessboard.

TEXTBOX 14.1:   The Fable of the Rice and the Chessboard

"The story goes that the ruler of India was so pleased with one of his palace wise men, who had invented the game of chess, that he offered this wise man a reward of his own choosing.

The wise man, who was also a wise mathematician, told his Master that he would like just one grain of rice on the first square of the chess board, double that number of grains of rice on the second square, and so on: double the number of grains of rice on each of the next 62 squares on the chess board.

This seemed to the ruler to be a modest request, so he called for his servants to bring the rice. How surprised he was to find that the rice quickly covered the chessboard, then filled the palace! Let's stop here, and see just how many grains of rice this is.

The number of grains of rice on the last square can be written as '2 to the 63th power', which looks like this: 2^{63}, which can be written as approximately: 18 446 744 070 000 000 000

A grain of rice is approximately 0.2 inches long. Converting 0.2 inches to feet (divide by 12 inches to a foot) and then dividing that number by 5,280 feet in one mile, we get the length of the grains of rice, placed end-to-end, to be approximately 60 000 000 000 000 miles. How far is that? Alpha Centauri, the nearest star, is located 25 000 000 000 000 miles from Earth. Placed end to end, these grains

of rice would reach farther than from the Earth, across space to the nearest star, Alpha Centauri, and back to Earth again!" Sanders, n.d.
   http://mathforum.org/sanders/geometry/GP11Fable.html
accessed Oct 2014.
   As an exercise, draw a causal loop diagram of this fable.

Systems dominated by balancing loops exhibit two types of behavior: 1) Gradual approach to a goal if they are first order systems – how gradual depends on the delay of the system or 2) oscillations around a goal if they are second or higher order systems.[1] If these oscillations are damped, sustained or exploding depends again on the delays of the system. In the latter case of exploding oscillations the system is not viable, sooner or later it ceases to exist. The goal is to reach an equilibrium, but reality usually intervenes so that the goal most of the time is over- or undershot.

The shipping system is thus characterized by damped oscillations of two wavelengths. A shorter 4 to 7 year oscillation controlled by the "capacity utilization" loop and a longer 20 year oscillation controlled by the "fleet capacity" loop. Their interaction creates the characteristic pattern of roughly 10 years of overcapacity, when freight rates are, from an investor's point of view, dismally flat and low and 10 years of undercapacity, when freights rates are generally high and reach 2 to 3 times during this period truly profitable heights.

The shipping system thus conceived constantly tries to balance transport supply to external demand. This is possible because the cost of transport is tiny compared to the value of what is being transported: the feedback in the form of freight rates has only a weak effect on the demand of transport. Or, in economic jargon: the price elasticity of transport demand is very low.

The inner loop via "fleet utilization" has a time horizon of 4 to 7 years[2] and adjusts all the levers of capacity limitation previously mentioned. The outer loop via ordering of new ships has a time horizon of about 20 years and

---

1   The order of a system is determined by the number of state variables, see http://en.wikipedia.org/wiki/State_variable or http://mathinsight.org/thread/elementary_dynamical_systems both accessed Oct 2014.

2   Time horizon of 4 to 7 years means a wavelength of 4 to 7 years. The delay is about one third of this wavelength.

adjusts the total stock of "fleet capacity." In this endogenous conceptualization "transport demand" has been relegated from a previously critical key variable, without which no sensible decision could be taken, to a mere trigger for the oscillations of the shipping system. This simple model is capable of generating the behavior over time of all the key variables of the system. (For details see Randers and Golüke 2007.)

WHAT HAS BEEN ACHIEVED SO FAR?

We have recast the problem for the ship investor from having to precisely forecast external demand for bulk transport – difficult if not impossible to do – to understanding the collective behavior of his or her peers – still difficult, but more doable.

14.3 MARKET SENTIMENT AND DETERMINISTIC BACKBONE

Jorgen coined an expression for "the collective behavior" of the shipping market participants; he calls it the "market sentiment," which captures the average mood of the shipping community. Every time this mood turns, opportunities for profit are created and having a reasoned judgment in advance of when the mood turns is, in fact, the value creating proposition of this modeling effort. Even when the mood is forecast *not* to turn, valuable information for the investor is generated about the timing of his or her investment decision.

Market sentiment generated by a *formal* model is a usable proxy for shipping investors because it already plays a prominent role in their *mental* models. The concept is useful when there is enough dynamic regularity in a system to overcome noise, which is a characteristic of any socio-economic system.

Jorgen also coined a second expression for this dynamic regularity: he calls it the "deterministic backbone" of the system. If that is strong enough, as it is in the case of investing in bulk shipping vessels, then the model can make useful predictions about the future of the market. Its momentum is strong enough to unfold into the next few years of the market, the very time frame in which investment decisions become profitable or not.[3]

---

3   What to do when the deterministic backbone is not strong enough? Then one has to redraw the system boundaries in such a manner that the deterministic backbone does indeed become strong enough. In fact, this advice is the essence of creating value through system dynamics models: draw the system boundaries in such a manner that the model's deterministic backbone is just strong enough to make and test predictions derived from the model. Tools to help you in this process are the traditional system dynamics tools of reference mode and the dynamic hypothesis. See (Randers 1980).

Knowing the deterministic backbone, we know both the patterns in the market and their respective strengths. To become the basis for decision making the only additional thing we need to know is where in the cycle we are and thus from which starting point the patterns unfold.

14.4 WHERE IN THE CYCLE ARE WE?

For investment purposes it suffices to make predictions over the coming 1 to 5 year time horizon. In other words, the model provides no value to the person in charge of negotiating freights, i.e. Chartering.[4] He or she lives in the 0 to 6 months time horizon. The investor, however, faced with the decision to order a new ship, purchase a secondhand one, lay up or scrap tonnage needs good information over the 1 to 5 year horizon. Forecasting the deterministic backbone of market behavior provides this reliable information.

As detailed by Randers and Golüke (2007), we worked long enough with the industry – over 30 years – to be able to create a track record of being able to predict these turning points 1 to 3 years ahead. Figure 10 in Randers and Golüke (2007) shows an example of such a track record. Since we began working on this in the 1970s we have seen the ten-year slump (oversupply) in the 80s and are currently (2014) again in such a period. Both the onset and the duration of both slumps were accurately forecast by the model.

FROM MODELING THE SYSTEM TO BUILDING UNDERSTANDING

In the end, the biggest challenge to creating value for shipping investors from a system dynamics model of their market turned out to be the presentation of results. As modelers we tended to be sure that the job was done once the model worked. That was due to *our* mental model, in this case of how people make decisions, which included an implicit and largely automatic translation of formal model representation and output to how we behave as a result of the mere existence of the formal model. It took us a long time to fully realize

---

4   "Chartering is an activity within the shipping industry. In some cases a charterer may own cargo and employ a shipbroker to find a ship to deliver the cargo for a certain price, called freight rate. Freight rates may be on a per-ton basis over a certain route (e.g., for iron ore between Brazil and China), in Worldscale points (in the case of oil tankers) or alternatively may be expressed in terms of a total sum – normally in US dollars – per day for the agreed duration of the charter.

A charterer may also be a party without a cargo who takes a vessel on charter for a specified period from the owner and then trades the ship to carry cargoes at a profit above the hire rate, or even makes a profit in a rising market by re-letting the ship out to other charterers." In addition to Wikipedia http://en.wikipedia.org/wiki/Chartering_%28shipping%29 accessed November 2014, see Zannetos 1966, Stopford 2000, Koopmans 1939 and Hageler 1977.

that formal models, even as simple (to us) as the one shown in the causal loop diagram above, simply do not exist for non-modelers. It is as if one assumes the existence of colors with all their nuances in a world of black and white. Communication across this gap is not only difficult, it does not occur – just as humans cannot really communicate with dogs about their respective senses of smell.

With hindsight, the solution looks easy – but then most intractable problems vanish with hindsight. In this case it was to modify (enhance?) ever so slightly the existing mental model of investors. What characterizes the mental model of investors? First and foremost, an uncanny ability to extract relevant patterns from oceans of detailed and often contradictory data. "Market sentiment" is one such key pattern, supported by much detail data, but still rising well above them. The second pillar of the mental models of shipping investors is freight rates – they are the only reliable real time data they have. Thus, the solution turned out to be as follows:

- Select a handful of freight rates to present as proxies for market sentiment.
- Show enough of their past to convey their deterministic backbone. Enough means in this case extending backwards over at least the last cycle of over and undersupply.
- Extend their momentum graphically 1 to 5 years into the future by splicing the forecast onto the average past.

As we summarize: "The resulting graph gives the reader, in one glance, both a quantitative reminder of the recent past and a picture of the likely future. … This splicing technique makes it possible to use far simpler models than would otherwise have been necessary. Without this technique we would need a more complicated model that could reproduce recent history within a few percent (to avoid disturbing jumps in the connection between history and forecast). A drawback is that such splicing may introduce quantitative inconsistencies in the forecast, for example if ordering and fleet size are scaled by different factors. But the time pattern remains uncorrupted" (Randers and Golüke 2007, p. 270).

This way of presenting fits easily into an investor's mental model. With a few slides, memorable images and storylines are created, and retained, that serve as critical background information when actually buying or selling vessels at the right time.

CONCLUSION

By developing three innovations, the deterministic backbone, the concept of market sentiment and the insight to splice the forecast onto average history, Jorgen and I were able to extend normal scientific system dynamics modeling concepts to provide real value over many decades to investors keen to profit from substantial opportunities in the bulk shipping market – an example of using science for (corporate) activism with a track record that by now has been proven.

While investment decisions are by nature confidential, we can report that one of the investors in our group, whose main activities were in the building sector, regularly realized significant profits from decisions based on our model advice. A second investor sold shortly before the onset of the slump starting in 2007–2008 his entire vessel portfolio to reinvest in solar energy. As forecast by our model at the time, the slump is still lasting. Timing, as the saying goes, is indeed everything.

BIBLIOGRAPHY

Bakken B. (1993). Learning and transfer of understanding in dynamic decision environments. PhD thesis. Cambridge, MA: MIT Sloan School of Management.

Hageler A. (1977). *En Analyse av Tankmarkedet i Perioden siden 1950 [A Study of the Tanker Market from 1950]*. Handelshøyskolen i København: Copenhagen

Koopmans, T.C. (1939). *Tanker Freight Rates and Tankship Building: An Analysis of Cyclical Fluctuations*. Haarlem, The Netherlands: PS King & Staple.

Randers, J. (1980). *Elements of the System Dynamics Method*. Cambridge, MA: MIT Press.

Randers, J. and Golüke, U. (2007). Forecasting turning points in shipping freight rates: lessons from 30 years of practical effort, *System Dynamics Review, 23*(2-3), 253–284, Summer–Autumn (Fall) 2007, DOI: 10.1002/sdr.376

Sanders, C.V. (no date). *The Geometry Pages*, accessible at http://mathforum.org/sanders/geometry/

Stopford, M. (2000). *Maritime Economics*. London: Routledge.

Taylor, A.J. (1976). System dynamics in shipping. *Operational Research Quarterly, 27*(1,i), 41–56.

Taylor, A.J. (1982). Chartering strategies for shipping companies. *OMEGA, 10*(1), 41.

Zannetos, Z.S. (1966). *The Theory of Oil Tankship Rates: An Economic Analysis of Tankship Operations*. Cambridge, MA: MIT Press.

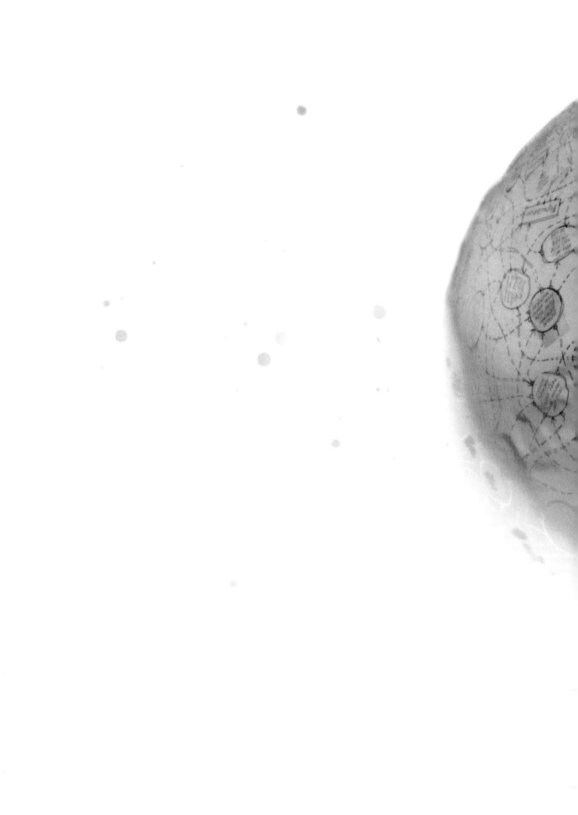

CHAPTER 15

Is Increased Use of Biofuels from Forests a Good Idea?

By Bjart Holtsmark

Traditionally, wood fuels, like other bioenergy sources, have been considered carbon neutral because the amount of CO_2 released can be offset by CO_2 sequestration due to regrowth of the biomass. Thus, emissions of biogenic CO_2 are usually not incorporated in carbon tax and emissions trading schemes. However, there is now increasing awareness of the inadequacy of this way of treating bioenergy, particularly bioenergy from the slow-growing boreal forests. This chapter considers the climate impact of increased use of bioenergy from slow-growing boreal forests. The conclusion is that such bioenergy has a larger climate impact in a 100-year time frame than fossil oil.

INTRODUCTION

As a response to the accumulation of CO_2 in the atmosphere, numerous countries have implemented subsidies to increase the uses of bioenergy. The argument is that combustion of bioenergy is "climate neutral" or "carbon neutral" because the harvest of one crop is replaced by the growth of a new crop, which reabsorbs the mass of carbon that was released by burning the first crop. This is a sensible argument in the case of annual plant crop-based biofuels, as new crops replace those that are harvested usually within one year. However, the carbon neutrality of the use of food-crop-based biofuels has been questioned. Fargione et al. (2008) found that converting new, native habitats to cropland releases CO_2 from both existing vegetation and carbon stored in soils. Fargione et al. (2008) therefore concluded that production of crop-based biofuels might create a biofuel carbon debt by releasing CO_2 at a level that is several times the level of annual greenhouse gas reductions that these biofuels would provide by displacing fossil fuels.

Searchinger et al. (2008) analysed the global effects of using grain or existing cropland for biofuel production. They argued that most previous analyses failed to take account of the carbon emissions that occur as farmers worldwide respond to higher crop prices and convert forest and grassland to new cropland to replace the grain or cropland diverted to biofuels.[1]

More generally, Wise et al. (2009) and Searchinger et al. (2009) underlined that the current practice of accounting for CO_2 emissions from combustion of bioenergy as zero means there are strong incentives to clear land, thus releasing large amounts of greenhouse gases.

The criticism of food-crop-based biofuels has not been directed toward wood-based biofuels to the same degree, at least not wood fuels from boreal

---

[1]   See also Gibbs et al. (2010), Gurgel et al. (2007), Lapola et al. (2010), and Melillo et al. (2009), among others.

forests. Even within the research community it has been common to consider timber from boreal forests as a carbon-neutral energy source; see for example Petersen and Solberg (2005), Raymer (2006), and Sjølie et al. (2010). Especially, the possibility of producing liquid biofuels from cellulosic biomass (second-generation biofuels) is considered a promising alternative to using food crops. For example, Killingland et al. (2013) studied the use of very large scale use of biomass from Norwegian forests for production of aviation fuel. They concluded that this is a promising sustainable option. However, their conclusion is based on the neutrality assumption mentioned above; i.e. that the emissions of CO_2 from combustion of the biomass could be ignored.

It would be reasonable to argue that wood fuels are carbon neutral if new trees grew so fast that they replaced those that are felled a year later. However, this is not the case in a boreal forest. Even after 10 or 20 years, new trees are still only saplings. In typical boreal-forested areas, it usually takes 70–120 years before a stand of trees is mature (Storaunet and Rolstad 2002). As will be shown, this long growth period implies that a higher level of harvest entails a lower stock of carbon stored in the forest. Hence, the assumption that wood fuels from boreal forests are climate neutral should be replaced with more realistic assumptions about the dynamic consequences of harvest in a boreal forest. At the same time, it should be taken into account that clear-cutting in boreal forest will temporarily increase the albedo, especially if there is a long season with snow cover in the considered area. Increased albedo will have a cooling effect.

This chapter also includes another aspect of bioenergy policies. It is often taken for granted that increasing the use of bioenergy will always lead to an immediate corresponding reduction in the use of fossil energy. This is too hasty a conclusion. To what extent increased use of bioenergy is reducing the use of fossil fuels is an uncertain matter and depends on how the bioenergy policy is designed. Therefore, this chapter does not take for granted that increasing the use of bioenergy leads to a corresponding reduction in the use of fossil fuels.

This chapter will have the main focus on bioenergy from the boreal forests. When analysing this issue, it should also be kept in mind that the boreal forest region stores approximately twice as much carbon as the tropical forest region and is thus the most important terrestrial carbon storage, see Figure 15.1. Moreover, the boreal forest is the home of some of the last intact terrestrial and aquatic ecosystems and large and diverse populations of mammals and birds.

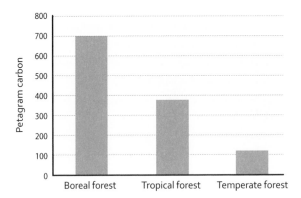

FIGURE 15.1: Carbon storage by global boreal, tropical and temperate forests. Source: Kasischke 2000.

15.1 A TRANSPARENT MODEL OF A BOREAL FOREST

The basis for any assessment of the climatic effects of bioenergy from a forest is a comparison of a no-harvest scenario and a harvest scenario. For both scenarios we need the time profile of the forest's total carbon stock and the corresponding net fluxes of CO_2 between the forest and the atmosphere.

To make things simple and transparent, I will in this chapter consider a stylized forest model consisting of 100 stands, each with properties as a typical Norwegian spruce forest. The stands are considered mature and ready for harvest at a stand age of 100 years. It is assumed that the stands' ages (years since last harvest) varies as follows. The age of stand #1 is 100 years in year $t = 0$ and it is ready for harvesting. The age of stand #2 is 99 years at time $t = 0$, and it will thus be ready for harvesting in year $t = 1$, and so forth. Hence, in year $t = 99$ the last stand is ready for harvesting. If we consider the harvest scenario, stand #1 will again be mature and ready for harvesting in year $t = 100$, and a complete new rotation will follow over the next century.

Figure 15.2 describes the development of the carbon pools of stand #1 in both the no-harvest-scenario (left diagram) and the harvest-scenario (right diagram). Either the stand is harvested (clear-cutting) at time $t = 0$ and $t = 100$, or the stand is not harvested.

The growth function for the trees on the stand was calibrated to fit into the standard production tables for Norway spruce of medium productivity provided in Braastad (1975). The volume of trunks is 194 m³/ha at the stand age of 100 years (Eid and Hobbelstad 2000). At the time of harvest ($t = 0$), the stand has a total carbon stock of 156 tC/ha (before harvesting,). In the harvest scenario, all stems of living trees are removed from the stand at time $t = 0$,

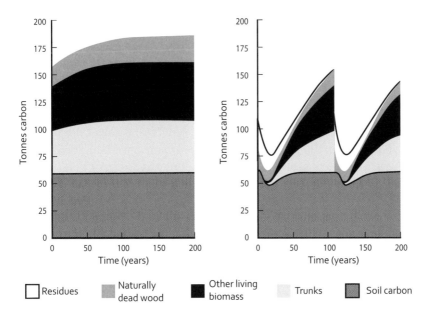

Residues ☐ Naturally dead wood Other living biomass Trunks Soil carbon

FIGURE 15.2: The development of carbon stored in a stand as modelled in the present article. The left diagram shows the no-harvest scenario while the left diagram shows the harvest scenario.

with subsequent combustion giving rise to a pulse of CO_2 corresponding to the amount of carbon contained in the stems (39 tC/ha) and the collected residues (75 per cent of tops and branches, 11 tC/ha). Hence, after harvesting at time $t = 0$, the stand stores 106 tC/ha (including soil carbon, residues left on the forest floor, and naturally dead wood); see Figure 15.2.[2]

New trees start growing after harvesting; see the hatched and cross-hatched areas in the diagram to the right of Figure 15.2. Residues left on the forest floor decompose; see the black area. Moreover, natural deadwood (NDOM) that was present in the stand at the time of harvesting also gradually decomposes, while new, naturally dead biomass is generated; see the dotted areas in Figure 15.2.

---

2   Note that the numerical results could paint too optimistic a picture of the use of residues for energy purposes, as it was assumed that the removal of residues does not influence future growth or the release of soil carbon (Helmisaari et al. 2011, Palosuo et al. 2001). It is taken into account that the use of residues is resulting in a correspondingly higher emission pulse at time t = 0 and correspondingly smaller subsequent emissions from the decomposition of residues (Repo et al. 2011, 2012).

As regards the dynamics of the soil carbon pool, it was assumed that harvesting results in some years with a net release of carbon from the soil (Buchholz et al. 2013, de Wit and Kvindesland 1999). Thereafter, the soil carbon pool gradually returns to its original state; see Figure 15.2, diagram to the right.

The development of the stand's carbon stock in the no-harvest baseline scenario is shown in Figure 15.2, diagram at left. The point of departure is that the stand's age is 100 years at $t = 0$. Hence, at time $t = 0$ in the no-harvest scenario, the sizes of the carbon pools are the same as at time $t = 100$ in the harvest scenario, cf. Figure 15.2. Moreover, in the no-harvest scenario, there is continued forest growth after $t = 0$, with a corresponding continued accumulation of natural deadwood (Faustmann 1849, Scorgie and Kennedy 1999). In the no-harvest scenario, the soil's carbon pool is assumed to be constant over time. This is a simplification. The soil carbon stock is increasing over time in boreal forests.

There is significant uncertainty about the likely development of the carbon stock of an old stand. However, in line with, e.g., Luyssaert et al. (2008) and Carey et al. (2001), I assumed continued accumulation of carbon even in old stands, although this accumulation is assumed to decrease toward zero over time. As this is an uncertain part of the scenario, Holtsmark (2014a) provided a sensitivity analysis with a significantly smaller accumulation of carbon in older stands.

Further details about all assumptions made are given in Holtsmark (2014a).

ACCUMULATION OF CARBON IN
THE ATMOSPHERE AND RADIATIVE FORCING EFFECTS

To estimate the climatic consequences of bioenergy and fossil fuels, it is beneficial to take into consideration the lifetime of atmospheric CO_2. A pulse of CO_2 into the atmosphere leads to increased absorption of CO_2 by the terrestrial biosphere as well as the sea. The Bern 2.5CC carbon cycle model was applied to take such dynamic effects into account (Joos and Bruno 1996, Joos et al. 1996, and Joos et al. 2001). The profile of the broken curve in Figure 15.3 depicts this model's prediction of the remaining proportion at time t of a CO_2 pulse generated at time $t = 0$. The Bern 2.5CC model is also applied to both the pulse emission caused by combustion of the harvest and to the fluxes of CO_2 generated by the stands' growth and by the decomposition of natural deadwood and harvest residues left on the forest floor; see Holtsmark (2014a) for further details.

To make the potential warming effect of CO_2 emissions comparable to the cooling effect of increased albedo, it is necessary to model the additional radiative forcing of additional carbon in the atmosphere. It was here assumed that the climate sensitivity of CO_2 is 3°C.

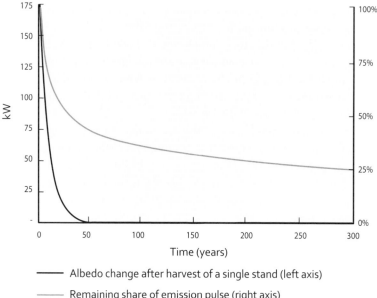

Albedo change after harvest of a single stand (left axis)

Remaining share of emission pulse (right axis)

FIGURE 15.3:   The broken curve shows the remaining share at time t of an emission pulse of CO_2 at time 0. The solid curve shows the increased albedo from a single stand that is harvested at time t = 0 and how the albedo effect is gradually reduced as regrowth takes place. The axis at left shows the change of the top-of-the-atmosphere albedo measured in kW. Given that the surface of the planet is approximately $5.10072 \times 10^{14} \, m^2$, 1 kW in absolute terms corresponds to $1.9605 \times 10^{-14} \, kW/m^2$ top-of-the-atmosphere albedo change.

Next, consider the albedo effect. It was assumed that clear-cutting of a stand of 1 ha results in an immediate rise in albedo. The albedo effect is gradually reduced as regrowth takes place; see the broken curve in Figure 15.3. In practice, the albedo effect from clear-cutting a site varies significantly depending on the exact site considered. The calculation of the albedo effect was here based on the model applied in Holtsmark (2014a), but with a somewhat more rapidly decreasing albedo effect after harvest, as the assumptions made in Holtsmark (2014a) indicate a significantly higher albedo effect even after 50 years, which appears unrealistic.

CALCULATION OF CO2 EMISSIONS FROM THE COMBUSTION OF BIOMASS

As mentioned, harvesting at the stand age of 100 years was assumed to yield 194 m^3 of wood per hectare when only the trunks were harvested. In addition, approximately 75 per cent of tops and branches were harvested together with the stems. According to Holtsmark (2014a), tops and branches are assumed to constitute 17 per cent of the trees' total biomass (including both above and below ground). Hence, the collected harvest residues are assumed to amount to 53 m^3/ha, and a total of 247 m^3/ha of wood is harvested in that case. As 1 m^3 of wood is assumed to contain 200 kg C, this means that combustion of the harvest releases (247 x 200 =) 49.3 tC/ha.

The assumption made here, i.e. that the entire harvest is used for energy purposes, is common in the literature, but it should nevertheless be discussed. Typically, 20–30 per cent of the stems are used as building materials or furniture. Moreover, a large proportion of the harvest is usually used as input in the pulp and paper industry. Although almost all the biomass that is used in the pulp and paper industry can be assumed to be combusted within less than a year after harvesting, it might seem unrealistic to study a case where 100 per cent of the biomass is used for energy purposes.

There are reasons for this choice, however. The point of departure for the analysis is a large-scale increased harvesting aimed at expanding the supply of biomass for energy purposes. Different kinds of subsidies are likely tools for achieving such expansion, and a number of subsidies are already in place. For example, the fact that CO_2 emissions from bioenergy are not included in the EU-ETS is an implicit subsidy that has effect in Europe. These subsidies lead to increased demand for wood fuels, followed by higher wood prices and thereby increased harvesting. However, there are few reasons to believe that increased harvesting will lead to increased use of wood in buildings and furniture, etc. Instead, higher wood prices as a result of the above-mentioned policies promoting increased use of bioenergy are likely to decrease the use of wood in buildings. If the harvest level increases while, at the same time, demand for biomass for construction purposes remains unchanged, or possibly decreases, it is appropriate to make the assumption that the whole harvest is used for energy purposes. It is essential here to keep in mind that the purpose of the study is not to analyse the climatic effects of forestry as such, but rather the climatic effects of increased harvesting to increase the supply of bioenergy.

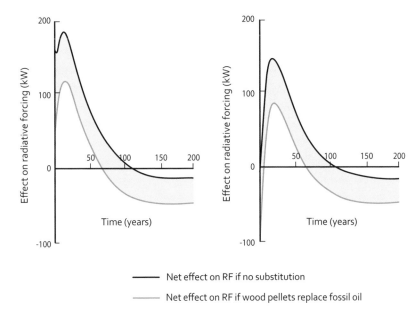

———— Net effect on RF if no substitution

———— Net effect on RF if wood pellets replace fossil oil

FIGURE 15.4:   A single harvest event. The effect on radiative forcing (RF) if a single stand of age 100 years is harvested in year t = 0. The left-hand diagram shows the case without any albedo changes, while a significant increase in albedo is assumed in the right-hand diagram. The double-lined curves show the net effect of harvesting on RF if an increased supply of bioenergy does not lead to any reduction in the consumption of fossil fuels. The single-lined broken curves show the limiting case where each kWh of wood fuels replaces one kWh of fossil oil.

15.2 RESULTS

The previous section introduced the model of a forest of 100 identical stands, each of 1 ha. This model will in the following be used to study the climatic effects of harvesting to produce bioenergy.

Before the model of the entire forest is applied, I present for illustrative purposes the climatic consequences of harvesting stand # 1 in year $t = 0$ only, see Figure 15.4. The double-lined curves show the time profile of the net effect on RF of harvesting this stand if the increased amount of bioenergy does not lead to any reduction in the consumption of fossil fuels. In the diagram at left, albedo effects are not considered, while the diagram at right has included the effects of increased albedo after harvesting. In both cases we see that harvesting leads to increased RF for more than 100 years after harvesting (in the case with enhanced albedo, there will be a very short initial period with reduced

RF). After approximately 110 years the double-lined curve drops below zero, which means a net cooling effect.

Assuming that increased supply of bioenergy does not lead to less fossil fuel consumption is a limiting and unlikely case. It is more likely that increased supply of bioenergy will lead to lower energy prices, which, in turn, will lead to reduced supply, and thus also to reduced consumption of fossil fuels. It will here be assumed that wood is a raw material used for wood pellets production and that the wood pellets replace liquid oil consumption. This means less demand for oil. To what extent lower demand for oil at the end of the day means lower global oil consumption is difficult to estimate because it depends on how global oil supply is changed both in the short and long term as a result of reduced demand. The difficult task of estimating supply and demand in the energy markets goes beyond the limits of this chapter. However, the chapter endeavours to provide information about the degree of uncertainty at this point by reporting both the limiting case where 1 kWh of wood fuels replaces 1 kWh fossil oil and the limiting case with no substitution of fossil fuels at all, as described above.[3] The most likely outcome is somewhere in between these two limiting cases and is therefore illustrated by the shaded areas in Figure 15.4.

Comparing the right and left-hand diagrams of Figure 15.4, we see that the results are identical in the long term, but different in the short term. The differences in the short term appear because the right-hand diagram assumes a significant albedo effect of harvesting, while the left-hand diagram illustrates the case without any albedo changes. In the albedo case, an immediate net cooling effect was assumed. A period of net warming follows, also when substitution is taken into account. However, 55–110 years after the harvesting, there will again be a net cooling effect because the stand has regrown and absorbed the carbon that was released at the time of harvesting. Note that there will be a cooling effect in the long term even in the case with no substitution of fossil fuels. This is because part of the released biogenic carbon has been absorbed by the sea and the terrestrial biosphere. Hence, in the long term, the net effect of the single harvest event is reduced CO_2 concentration

---

3   The substitution is based on Holtsmark (2012), where it was assumed that the energy from 1 m³ wood could eliminate 0.5 tCO_2 from a fossil source. This is consistent with the pellets case considered in Holtsmark (2012), where pellets replace coal. If the pellets instead replace oil, then 1 m³ wood used as energy could replace 0.72 x 0.5 = 0.36 tCO_2 from a fossil source. This is the same as 0.136 tC. As 1 m³ contains 0.2 tC, this means that we have a factor of substitution of 0.136/0.2 = 0.6818.

in the atmosphere, also in the case where the increased supply of wood fuels does not replace any fossil fuel consumption.

To capture the true effects of increased use of bioenergy, it is not sufficient to consider a single harvest at time $t = 0$, as this will mean some supply of bioenergy at time t = 0, but nothing in later periods. Hence, a case with supply of bioenergy every year should be considered. Therefore, I now turn to the multiple harvest (landscape) approach. This means that in the harvest scenario every year a new stand will be harvested. The time profiles of the carbon stock of the stands are illustrated by the diagram to the right of Figure 15.2, except that there will be time offsets corresponding to the stands' different ages.

Figure 15.5 shows how the carbon stock of *the entire forest* develops in both scenarios. In the harvest scenario the carbon stock will be constant over time, while the carbon stock will be increasing in the no-harvesting-scenario before it is stabilized at a higher level than in the harvesting-scenario. Hence, bioenergy should not even in the long term be considered as carbon neutral.

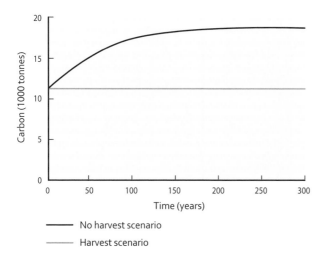

FIGURE 15.5: The carbon stock of the considered forest in the harvest-scenario and the no-harvest-scenario.

Figure 15.6 shows the results of the multiple harvesting approach with regard to the net effect on RF. Again, the right-hand diagram is based on a significant albedo effect of harvesting, while the left-hand diagram does not include any albedo changes.

First, note that using the multiple harvesting (landscape) approach leads to a significantly different outcome compared to the single harvest (stand level) approach. Without substitution of fossil fuels (the double-lined curves), there is a warming effect in the whole time span shown (200 years), not only temporary as in the single harvest case. As also reported in Holtsmark (2013), Figure 15.3, there will be a permanent warming effect if no substitution of fossil fuels takes place. This should be seen in relation to Figure 15.5 and the permanently lower carbon stock of the forest in the harvesting-scenario compared to the no-harvesting-scenario.

When there is substitution of fossil fuels, the results are different. First, consider the case without albedo effects (the left-hand diagram of Figure 15.6). If we consider the limiting and very optimistic case with full substitution in the sense that 1 kWh of wood pellets replaces 1 kWh of oil (the broken curve), there will be an initial period of 180 years with net warming. This time period is often called "the payback time of the carbon debt". After these 180 years there will be reduced RF.

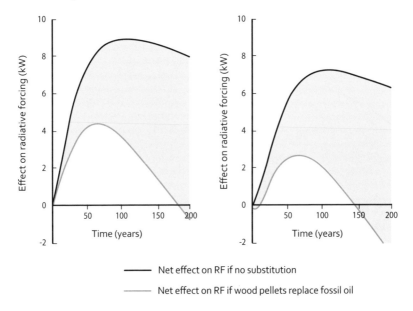

FIGURE 15.6: The multiple harvest (landscape) approach. The effect on RF if, every year from t = 0, one stand 100 years of age is harvested. It was assumed that most (75 per cent) of the tops and branches were harvested together with the stems. The entire harvest was used for bioenergy. The left-hand diagram shows the case without any albedo changes, while a significant increase in albedo is assumed in the right-hand diagram. The double-lined curves show the net effect of harvesting on RF if an increased supply of bioenergy does not lead to any reduction in the consumption of fossil fuels. The single-lined broken curves show the limiting case where each kWh of wood fuels replaces one kWh of fossil oil.

Next, consider the case with albedo effects (the right-hand diagram in Fig. 6). In that case, there will be an immediate cooling effect after time $t = 0$, at least if there is full substitution. However, this cooling effect is temporary. After 12 years there will be a net warming effect also in the case with full substitution and this warming period will last for approximately 140 years. However, that is only the case with full substitution on a 1 kWh against 1 kWh basis. It is more likely that the degree of substitution will be smaller, leading to a less favourable effect of harvesting that lies somewhere in the grey area between the broken and double-lined curves.

DISCUSSION AND CONCLUSION

To provide a picture of the likely climatic effects of increased harvesting of a boreal forest using bioenergy to replace fossil fuels, the previous section presented simulations with a model of a typical slow-growing boreal forest. The model of the forest stand is not a "black box", but rather a transparent model that easily could be understood and checked in all its details by the reader. The model forest includes 100 stands, all with the same growth path after clear-cutting and replanting, although the stands' ages since last felling are distributed evenly such that every year a new stand will reach the stand age of 100 years. Figure 15.2 describes the model of the individual stands. To estimate the climatic consequences of bioenergy and fossil fuels, the forest model was linked to the Bern 2.5CC carbon cycle model. This model incorporates how a pulse of CO_2 into the atmosphere leads to increased absorption of CO_2 by the terrestrial biosphere as well as the sea. And finally, a model for the albedo effect of harvesting was incorporated.

Before the results are discussed, some limitations and characteristics of the present study should be emphasized. What is most important to keep in mind is that the study considers additional harvesting solely for the purpose of increasing the supply of bioenergy. And the study considers only one (the climatic) out of many aspects of forestry. Hence, the study does not represent an evaluation of forestry in general. Moreover, the use of leftover biomass from the forest industry, which otherwise had not been used for energy purposes, is not analysed in this study. The assumption that the entire harvest is used for energy purposes was discussed at the end of the Model-section.

Note also that the present study does not discuss the size of the albedo effects. At this point, the assumptions are fully based on Cherubini et al. (2012), who considered a forest stand located in Hedmark in the south-eastern part of Norway. This is an area with a relatively long season with snow cover. Many other districts in Norway have either shorter snow seasons or are located at higher altitudes with winters with less sun. Hence, most forests in Norway will probably have smaller albedo effects of harvesting.

Moreover, the albedo effect of harvesting is assumed to be the same throughout the simulation period. This does not take into account that a warmer climate is likely to reduce the season with snow cover during the 21st century. Hence, the cooling effect of albedo after harvesting could have been overestimated in the present calculations.

It should also be kept in mind that there is a significant degree of uncertainty about the warming effect of the increasing concentration of CO_2. The simulations in this study are mainly based on the standard 3°C climate sensitivity assumption. If the climate sensitivity of CO_2 is higher, the albedo effect will be less important compared to the CO_2-effect of harvesting. Hence, from a precautionary perspective, it could be risky to carry out harvesting based on the assumption that the albedo effect will to some extent neutralize the warming effect of increased CO_2 in the atmosphere.

An uncertain element in the study is the accumulation of carbon in older stands. Readers concerned about this type of uncertainty are directed to Holtsmark (2014), who included a sensitivity analysis in that respect. This sensitivity study did not alter the results fundamentally, but gave results that were somewhat more in favour of bioenergy.

The present study does not include a scenario in which the trees on the replanted stand show higher productivity than the original trees. However, the supplementary online material in Holtsmark (2012) included that type of scenario and showed how the payback time of the carbon debt was then reduced. But still the results were not fundamentally changed.

Some words should be included about the results and how they should be interpreted. Figure 15.5 shows how the forest's carbon stock develops in both the harvesting and the no-harvesting scenarios. The noteworthy feature here is that in the long run the forest's carbon stock is lower in the harvest scenario than in the no-harvest scenario. Compared with Figure 15.2, this result appears obvious, as the studied stands' carbon stocks at any point in time are lower in the harvest scenario compared to the no-harvest scenario. It follows that it is misleading to consider bioenergy from a slow-growing forest as a climate neutral or carbon neutral energy source. In neither the short nor the long term harvesting are a slow-growing forest and combustion of the biomass carbon neutral activities.

Although the forest's carbon stock also in the long term is lower in the harvest scenario compared to the no-harvest scenario, bioenergy will have a net cooling effect in the long term if bioenergy replaces fossil fuel consumption. Note, however, that the concept "long term" here means centuries. Even with the most optimistic assumptions about how much fossil oil could be replaced, large scale use of bioenergy from boreal forests would mean increased global warming for the next 140–200 years. Hence, it is misleading to claim that the warming effect of bioenergy from a boreal forest is a short term effect.

The point where the broken curve of the left-hand diagram of Figure 15.6 crosses the x-axis defines the length of what has been labelled the payback time of the carbon debt (Fargione et al. 2008, Dehue 2013, Holtsmark 2012, Jonker and Junginger 2013, Lamers et al. 2013, Lapola et al. 2010, Searchinger et al. 2008). The payback time of the carbon debt was here found to be 180 years if there is no cooling effect from increased albedo. If there is a cooling effect from albedo, there is an initial short period of net cooling before there is a longer period of net warming. Approximately 150 years after the bioenergy policy was initiated there will again be net cooling (Figure 15.6, right-hand diagram).

Finally, some words about the policy relevance of the results in this study. It was found that if bioenergy from boreal forests replaces fossil fuel consumption, there might be climate gains in the very long term, but only after at least a century with increased warming. In the case where bioenergy replaces gas, "short term" is here approximately 100–150 years, while it is less in the cases of coal and oil. This means that if the very long term is more important than a shorter time perspective, bioenergy might still be preferred to fossil fuels. If, on the other hand, there is concern that the climate is approaching a tipping point within a timeframe of less than 100 years, increased use of bioenergy from slow-growing forests could increase the climate risk.

BIBLIOGRAPHY

Braastad, H. (1975). *Yield Tables and Growth Models for Picea abies.* Reports from The Norwegian Forest Research Institute 31.9.

Buchholz, T., Friedland, A.J., Hornig, C.E. et al. (2013). Mineral soil carbon fluxes in forests and implications for carbon balance assessments. *Global Change Biology Bioenergy*, doi: 10.1111/gcbb.12044 (2014).

Carey, E.V., Sala, A., Keane, R., Callaway, R.M. (2001). Are old forests underestimated as global carbon sinks? *Global Change Biology*, 7, 339–344.

Cherubini, F., Bright, M., Strømman, A.H. (2012). Site-specific global warming potentials of biogenic CO2 for bioenergy: contributions from carbon fluxes and albedo dynamics. *Environmental Research Letters*, 7, doi:10.1088/1748-9326/7/4/045902.

de Wit, H.A., Kvindesland, S. (1999). *Carbon stocks in Norwegian forest soils and effect of forest management on carbon storage.* Report 19/1999 from Norwegian Forest and Landscape Institute.

Dehue, B. (2013). Implications of a 'carbon debt' on bioenergy's potential to mitigate climate change. *Biofuels, Bioproducts and Biorefining*, 7, 228–234.

Eid, T. and Hobbelstad, K. (2000). AVVIRK-2000 – a large scale scenario model for long-term investment, income and harvest analyses. *Scandinavian Journal of Forest Research*, 15, 472–482.

Fargione, J., Hill, J., Tilman, D., Polasky, S. and Hawthorne, P. (2008). Land clearing and the biofuel carbon debt. *Science*, 319, 1235–1238.

Faustmann, M. (1849). Berechnung des Werthes, weichen Waldboden so wie nach nicht haubare Holzbestande für de Weltwirtschaft besitzen. *Allgemeine Forst und Jagd Zeitung*, 25, 441.

Gibbs, H.K., Ruesch, A.S., Achard, F., Clayton, M.K., Holmgren, P., Ramankutty, N. and Foley, J.A. (2010). Tropical forests were the primary sources of new agricultural land in the 1980s and 1990s. *P NatlAcadSci*, 107:16732–16737.

Gurgel, A.J., Reilly, M. and Paltsev, S. (2007). Potential land use implications of a global biofuels industry. *J Agr Food Ind Organ*, 5, 1–34.

Helmisaari, H.-S., Hanssen, K.H., Jacobson, S. et al. (2011). Logging residue removal after thinning in nordic boreal forests: long-term impact on tree growth. *Forest Ecology and Management*, 261, 1919–1927.

Holtsmark, B. (2012). Harvesting in boreal forests and the biofuel carbon debt. *Climatic Change*, 112, 415–428.

Holtsmark, B. (2013). The outcome is in the assumptions: analyzing the effects on atmospheric CO2 levels of increased use of bioenergy from forest biomass. *Global Change Biology Bioenergy*, 5, 467–473.

Holtsmark, B. (2014a). Quantifying the global warming potential of CO_2 emissions from wood fuels. Forthcoming in *Global Change Biology Bioenergy* (DOI: 10.1111/gcbb.12110).

Holtsmark, B. (2014b). A comparison of the global warming impact of wood fuels and fossil fuels taking albedo effects into account. *GCB Bioenergy* (in press).

Jonker, J.G.G., Junginger, M., Faaij, A. (2013). Carbon payback period and carbon offset parity point of wood pellet production in the South-eastern United States. *Global Change Biology Bioenergy*, doi: 10.1111/gcbb.12056 (2014).

Joos, F., Bruno, M. (1996). Pulse response functions are cost-efficient tools to model the link between carbon emissions, atmospheric CO2 and global warming. *Physics and Chemistry of the Earth*, 21, 471–476.

Joos, F., Bruno, M., Fink, R., Stocker, T.F., Siegenthaler, U., Le Quere, C. and Sarmiento, J.L. (1996). An efficient and accurate representation of complex oceanic and biospheric models of anthropogenic carbon uptake. *Tellus*, *48B*, 397–417.

Joos, F., Prentice, I.C., Sitch, S. et al. (2001). Global warming feedbacks on terrestrial carbon uptake under the Intergovernmental Panel on Climate Change (IPCC) emission scenarios. *Global Biogeochemical Cycles*, *15*, 891–907.

Kasischke, E.S. (2000). Boreal Ecosystems in the Global Carbon Cycle. In E.S. Kasischke and B.J. Stocks (Eds.) *Fire, Climate and Carbon Cycling in the Boreal Forest*. New York: Springer.

Killingland, M., Aga, P., Grinde, M., Saunes, I., Melbye, A.M., Vevatne, J. and Christophersen, E.B. (2013). *Bærekraftig biodrivstoff for luftfart (Sustainable biofuels for aviation)*. Report from Rambøll Consulting.

Lamers, P. and Junginger, M. (2013). The 'debt' is in the detail: a synthesis of recent temporal forest carbon analyses on woody biomass for energy. *Biofuels, Bioproducts and Biorefining*, doi: 10.1002/bbb.1407 (2014).

Lapola, D., Schaldach, M.R., Alcamo, J., Bondeau, A., Koch, J., Koelking, C. and Priess, J.A. (2010). Indirect land-use changes can overcome carbon savings from biofuels in Brazil. *Proceedings for the National Academy of Sciences*, *103*, 11206–11210.

Luyssaert, S., Schulze, E.D., Börner, A., Knohl, A., Hessenmöller, D., Law, B.E., Ciais, P. and Grace, J. (2008). Old-growth forests as global carbon sinks. *Nature, 455*, 213–215.

Melillo, J.M., Reilly, J.M., Kicklighter, D.W., et al. (2009). Indirect emissions from biofuels: how important? *Science, 326*, 1397–1399.

Palosuo, T., Wihersaari, M., Liski, J. (2001). Net greenhouse gas emissions due to energy use of forest residues – impact of soil carbon balance. In *EFI Proceedings No. 39, Wood Biomass as an Energy Source Challenge in Europe*. European Forest Institute, Joensuu, pp. 115–130.

Petersen, A.K. and Solberg, B. (2005). Environmental and economic impacts of substitution between wood products and alternative materials: a review of micro-level analyses from Norway and Sweden. *Forest Policy Econ, 7*, 249–259.

Raymer, A.K.P. (2006). A comparison of avoided greenhouse gas emissions when using different kinds of wood energy. *Biomass Bioenerg, 30*, 605–617.

Repo, A., Kankanen, R., Tuovinen, J.P., Antikainen, R., Tuomi, M., Vanhala, P. and Liski, J. (2012). Forest bioenergy climate impact can be improved by allocating forest residue removal. *Global Change Biology Bioenergy*, *4*, 202–212.

Repo, A., Tuomi, M. and Liski, J. (2011). Indirect carbon dioxide emissions from producing bioenergy from forest harvest residues. *Global Change Biology Bioenergy*, *3*, 107–115.

Scorgie, M. and Kennedy, J. (1996). Who discovered the Faustmann Condition? *History of Political Economy*, *28*, 77–80.

Searchinger, T.D., Hamburg, S.P., Melillo, J. et al. (2009). Fixing a critical climate accounting error. *Science*, *326*, 527–528.

Searchinger, T.D., Heimlich, T.D., Houghton, R.A., et al. (2008). Use of US croplands for biofuels increases greenhouse gas through emissions from land-use change. *Science, 319*, 1238–1240.

Sjølie, H.K., Trømborg, E., Solberg, B., Bolkesjø, T.F. (2010). Effects and costs of policies to increase bioenergy use and reduce GHG emissions from heating in Norway, *Forest Policy and Economics*, *12*, 57–66.

Storaunet, K.O. and Rolstad, J. (2002). Time since death and fall of Norway spruce logs in old-growth and selectively cut boreal forest. *Can J For Res, 32*, 1801–1812.

Wise, M., Calvin, K., Thomson, A., Clarke, L., Bond-Lamberty, B., Sands, R., Smith, SJ., Janetos, A. and Edmonds, J. (2009). Implications of limiting CO_2 concentrations for land use and energy. *Science, 324*, 1183–1186.

Randers Bibliography, Chronological

Meadows, D.H., Meadows, D.L., Randers J. and Behrens, W.W. (1972). *The Limits to Growth: A Report for the Club of Rome's Project on the Predicament of Mankind*. New York: Universe Books. Available in 36 languages.

Meadows, D.L. and Randers, J. (1972). Adding the Time Dimension to Environmental Policy. *International Organization, 26*(2), 213–233. doi: 10.1017/S0020818300003301

Randers, J. (1972). The dynamics of environmental policy: DDT. *The International Journal of Environmental Studies, 4.*

Randers, J. and Meadows, D. (1972). The Carrying Capacity of the Globe. *Sloan Management Review (pre-1986), 13*(2), 11.

Randers, J. and Meadows, D.L. (1972). The dynamics of solid waste generation. *Technology Review,* April.

Meadows, D.H., et al. (1973). A response to Sussex. *Futures, 5*(1), 135–152. doi: 10.1016/0016-3287(73)90062-1

Randers, J. (1973). Conceptualizing Dynamic Models of Social Systems: Lessons from a Study of Social Change. Ph.D. dissertation, MIT Sloan School of Management, Cambridge, Mass.

Randers, J. (1973). The Dynamics of Solid Waste Generation. In D.H. Meadows and D.L. Meadows (Eds.), *Toward Global Equilibrium: Collected Papers* (pp. 165–211). Cambridge, MA: Wright-Allen Press.

Meadows, D.L., Randers, J. and others (1974). *Dynamics of Growth in a Finite World*. Cambridge, Mass.: Wright-Allen Press.

Randers, J. (1976). *A system dynamics study of the transition from ample to scarce wood resources*. Dartmouth: Dartmouth College.

Behrens, W.W. and Randers, J. (1978). Watch for the foothills: signalling the end to growth in a finite world In D.W. Orr and M.S. Soroos (Eds.),

Ecological Perspectives on World Order. Chapel Hill: The University of North Carolina Press.

Randers, J. (1978). How to stop industrial growth with minimal pain. *Technological Forecasting & Social Change, 11*(4), 371–382. doi: 10.1016/0040-1625(78)90019-7

Randers, J. (1978). How to be a useful builder of simulation models. In J. Rose (Ed.), *Current Topics in Cybernetics and Systems.* Berlin: Springer.

Randers, J., Stenberg, L. and Kalgraf, K. (1978). *Skognæringen i overgangsalderen: en analyse av overgangen fra raskt voksende til tilnærmet konstant produksjonsvolum i Nordens skognæring.* Oslo: Cappelen.

Ervik, L., et al. (1979). *Smelteverkene i smeltedigelen,* Oslo: Cappelen.

Lönnstedt, L. and Randers, J. (1979). *Wood resource dynamics in the Scandinavian forestry sector = Virkesbalansens dynamik i den skandinaviska skogsnäringen.* Uppsala: Swedish University of Agricultural Sciences, College of Forestry.

Randers, J. (1979). Norwegian petroleum policy and third world countries. *Jamaica Journal, 43,* 69–76.

Randers, J. and Lønnstedt, L. (1979). Transition strategies for the forest sector. *Futures, 11*(3), 195–205. doi: 10.1016/0016-3287(79)90109-5

Randers, J. (1980). *Elements of the system dynamics method.* Cambridge, Mass.: MIT Press.

Randers, J., Høsteland, J.E. and Lønnstedt, L. (1980). National strategies to limit the industrial use of wood to the sustainable forest growth. *Journal of Business Administration, 11*(1/2).

Randers, J. and Nilsson, J.W. (1980). Den unødvendige arbeidsløsheten. In O. Berrefjord et al. (Eds.), *Arbeid og inntekt.* Oslo: Tiden.

Randers, J. (1983). Guidelines for model conceptualization. In B.W. Hennestad and F.E. Wenstøp (Eds.), *Bedriftsøkonomi og vitenskapsteori: utvalgte tekster.* Oslo: Universitetsforlaget.

Randers, J. (1984). Prediction of pulp prices. A review two years later. *BI Working Paper 84/6*, Stiftelsen BI, Bekkestua.

Randers, J. (1985). The Tanker Market, *BI Working Paper 84/9*, Stiftelsen BI, Bekkestua.

Randers, J. et al. (1986). In OECD, *The public management of forestry projects.* Paris: OECD.

Randers, J. and Golüke, U. (1990–2009). Future Mapper for the Shipping Market. *Quarterly (later annual) market forecast to shipping investors.* Oslo: Future Mappers AS.

Meadows, D.H., Meadows, D.L. and Randers, J. (1992). *Beyond the limits: global collapse or a sustainable future.* London: Earthscan.

Meadows, D.H., Meadows, D.L. and Randers, J. (1992). *Beyond the limits: confronting global collapse, envisioning a sustainable future.* Post Mills, Vt.: Chelsea Green Publ. Co.

Randers, J. (1994). The quest for a sustainable society: a global perspective. In G. Skirbekk (Ed.), *The Notion of Sustainability and its normative implications* Oslo: Scandinavian University Press.

Randers, J. (1996). Depressing trends. In H. de Mattos-Shipley (Ed.), *Chaning worlds.* London: WWF.

Loh, J., Randers, J. et al. (1998–2004). *The Living Planet Report,* Gland: WWF International.

Randers, J. (2000). From limits to growth to sustainable development or SD (sustainable development) in a SD (system dynamics) perspective. *System Dynamics Review, 16*(3), 213–224.

Wackernagel, M., et al. (2002). Tracking the ecological overshoot of the human economy. *Proceedings of the National Acadademy of Sciences, 99*(14), 9266–9271. doi: 10.1073/pnas.142033699

Meadows, D.H., Randers, J. and Meadows, D. (2004). *Limits to growth: the 30-year update.* London: Earthscan.

Loh, J., et al. (2005). The Living Planet Index: using species population time series to track trends in biodiversity. *Philosophical Transactions of the Royal Society of London. Biological Sciences, 360,* 289–295.

Lærum, F., et al. (2005). Moving equipment, not patients: Mobile, net-based digital radiography to nursing home patients. *Excerpta Medica: International Congress Series, 1281,* 922–925. doi: http://dx.doi.org10.1016

Randers, J. (2006). Et klimavennlig Norge. In *NOU 2006:18.* Oslo: Departementene.

Isachsen, A.J. and Randers, J. (2007). Fremtidsbilde 2030: Er verden flat? Drivkrefter og spilleregler. *Magma – Tidsskrift for økonomi og ledelse, 10*(1), 51–57.

Randers, J. and Alfsen, K. (2007). How Can Norway Become a Climate-Friendly Society? *World Economics, 8*(1), 75–106.

Randers, J. and Golüke, U. (2007). Forecasting Turning Points in Shipping Freight Rates: Lessons from 30 Years of Practical Effort. *System Dynamics Review, 23*(2-3), 253–285.

Randers, J. (2008). Global Collapse – Fact or Fiction. *Futures: The journal of policy, planning and futures studies, 40*(10), 853–864. doi: http://dx.doi.org10.1016/j.futures.2008.07.042

Randers, J. and Gilding, P. (2010). The One Degree War Plan. *Journal of Global Responsibility, 1*(1), 170–188.

Ericson, T. and Randers, J.R. (2011). *A forward looking, actor based, indicator for climate gas emissions*. Oslo: CICERO.

Randers, J. (2011). Meny 5 – en vedtakbar klima- og energiplan for Norge til 2020. *Magma – Tidsskrift for økonomi og ledelse, 14*(2), 21–34.

Randers, J. (2012). Greenhouse gas emissions per unit of value added («GEVA»): A corporate guide to voluntary climate action. *Energy Policy, 48*, 46–55. doi: http://dx.doi.org10.1016/j.enpol.2012.04.041

Randers, J. (2012). 2052: Droht ein globaler Kollaps? *Aus Politik und Zeitgeschichte, 62*(51–52/2012), 3–10.

Randers, J. (2012). It's a Small World. *Foreign Affairs, 91*(5), 10–11.

Randers, J. (2012). The Real Message of The Limits to Growth A Plea for Forward-Looking Global Policy. *GAIA, 21*(2), 102–105.

Randers, J. (2012). *2052 – A Global Forecast for the Next Forty Years*. White River Junction Vt.: Chelsea Green Publishing. Available in 8 languages.

Randers, J. (2013). Global Trends 2030 Compared with the 2052 Global Forecast. *World Future Review, 5*(4), 360–366. doi: http://dx.doi.org10.1177/1946756713514543

Randers, J. (2013). Meeting the Climate Challenge: GEVA as an Aid to Corporate CSR. In A. Midttun (Ed.), *CSR and beyond: A Nordic perspective* (pp. 219–239). Oslo: Cappelen Damm Akademisk.

Rees, D., Marino, M.S. and Randers, J. (2013). *Norsk Klimapolitikk 2006–12* (pp. 33). Oslo: BIs Senter for Klimastrategi.

Rees, D. and Randers, J. (2013). *The «perceived progress» indicator – measuring the rate of change in subjective well-being* (pp. 13). Oslo: BIs Senter for Klimastrategi.

Randers, J. (2014). A realistic leverage point for one-planet living: more compulsory vacation in the rich world. *System Dynamics Review*, In press. doi: http://dx.doi.org10.1002/sdr.1522

Randers, J. (2014). Compulsory Vacation: Reducing the Human Ecological Footprint Through More Annual Leave *Disrupting the Future: Great Ideas for Creating a Much better World* (pp. 61–67). London: Kaleidoscope Futures.

Randers, J. (2014). Living in overshoot: A forecast and the desire to have it wrong. In *The colours of energy: Essays on the future of our energy system* (pp. 56–60). Amsterdam: Shell International BV.

Randers, J. (2014). 2052 – Japan: a global world leader in increasing citizen's well-being during slow GDP growth and declining population. *Fole, 2*, 2–5.

Randers, J. (2015). *Demokratin oförmögen att hantera klimahotet* (Can democracy handle difficult issues like climate change?). Extract, January 2015, Stockholm.

About the authors

Iulie Aslaksen is a senior researcher at the research department of Statistics Norway, working in the areas of sustainable development indicators, biodiversity policy, the Nature Index, and ecosystem services. With broad research experience ranging from oil price models to feminist economics, she explores ecological economics as an interdisciplinary approach to economics. She was a member of the Norwegian expert committee on values of ecosystem services (NOU 2013:10).

Hanne Inger Bjurstrøm is Norway's Special Envoy for Climate, Ministry of Climate and Environment. Bjurstrøm was Norway's Chief Climate Negotiator to the UNFCC from 2007–2010. From 2010–2012 she was Minister of Labor in the Stoltenberg II government.

Caroline Dale Ditlev-Simonsen holds a Ph.D. in Leadership and Organization (BI Norwegian Business School). Ditlev-Simonsen has international and comprehensive business and organizational experience in the areas of corporate responsibility, corporate citizenship and environmental and ethical issues. She has varied board experience and was a board member of WWF-Norway (World Wide Fund for Nature) from 2002–2008. Ditlev-Simonsen is also Co-Director at the BI Norwegian Business School's Centre for Corporate Responsibility www.bi.no/ccr

Kjell A. Eliassen is a Professor of Public Management and Director of the Centre for European and Asian Studies at BI Norwegian Business School and Professor of European Studies at The Free University in Brussels. He has been a professor at the University of Aarhus, Denmark and a visiting professor at several European, American and Asian Universities. His most recent books are: *Business and Politics in a Globalised World* (2012), *Analyzing European Union Politics* (2012), *Understanding Public Management* (2008), *European Telecommunications Privatisation* (2007).

John Elkington is co-founder and Executive Chairman of Volans. He also co-founded Sustain Ability (where he is Honorary Chairman) in 1987 and

Environmental Data Services (ENDS) in 1978. His latest book, *The Break-through Challenge: How to Connect Today's Profits With Tomorrow's Bottom Line*, is co-authored with Jochen Zeitz (former Chairman and CEO of PUMA and now co-Chair of The B Team) and will be published by Jossey-Bass this month. He thanks Astrid Hvam Høgsted for her help in preparing his essay.

Per Arild Garnåsjordet is a senior researcher at the research department of Statistics Norway, working in the areas of sustainable development indicators, biodiversity policy, the Nature Index, and ecosystem services. In the 1980s he chaired development of environmental statistics and natural resource accounting at Statistics Norway. He has been CEO of several Norwegian industrial and consulting companies before returning to research, initiating projects on knowledge for policy for sustainable development. He has contributed to the UN effort for developing experimental ecosystem accounting.

Paul Gilding has spent 35 years trying to change the world, doing everything he can think of. Despite the clear lack of progress, the unstoppable and flexible optimist is now an author and advocate, writing his widely acclaimed book *The Great Disruption*, which prompted Tom Friedman to write in the NYT: "Ignore Gilding at your peril." He now travels the world alerting people – in business, community groups, government and even the military – to the global economic and ecological crisis now unfolding around us, as the world economy reaches and passes the limits to growth.

Ulrich Golüke, born and raised in the Rhineland of Germany, studied systems dynamics with Dennis Meadows, co-author of the *"Limits to Growth"* study. He has been building feedback models ever since. For the last 20 years or so he has in addition worked with scenarios, as a way to imagine and create futures different than the official ones. He is a freelancer who works with companies, universities, foundations and students. He has lived in Wales, the United States, Norway, France and Switzerland. More info on www.blue-way.net.

Bjart Holtsmark is an economist (cand.oecon) and senior researcher at Statistics Norway. His research field is mainly environmental and resource economics. He is also affiliated with the Oslo Centre for Research on Environmentally friendly Energy (CREE). Among Holtsmark's latest publications are articles on environmental science and policy ("The Norwegian support and subsidy policy of electric cars," 2014) and in *The Journal of Forest Economics* ("Faustmann and the climate," 2014)

Gudmund Hernes is a researcher at the Fafo Institute in Oslo, and Professor II at BI Norwegian Business School. Gudmund Hernes was Norway's Minister of Education and Research 1990-95, and Minister of Health 1995-97. From 1999-2005 he was the Director of UNESCO's International Institute of Educational Planning in Paris, and UNESCO's Coordinator on HIV/AIDS. Gudmund Hernes has written and edited many specialized books, reports and, articles; he has been on the editorial board of several international specialist journals and is a member of several professional societies and academies. He was president of the International Social Science Council 2007-2010.

Jonathan Loh is a conservation biologist specializing in the monitoring and assessment of biological and cultural diversity. He is an Honorary Research Associate at the Zoological Society of London. He has developed indicators and monitoring systems for organizations such as WWF, the CBD and the Ramsar Convention.

Victor D. Norman is a Professor of Economics at NHH Norwegian School of Economics. He has a Ph.D. from MIT (1972). His special field is theory of international trade, but he has also published in a number of other areas, including economic growth, industrial organization, and "new" economic geography. He was rector at NHH from 1999 to 2001, and was a cabinet minister in the Norwegian government from 2001 to 2004.

Pavlina Peneva is currently a researcher on EU policy at Knowledge Development Ltd. Her main fields of research include public affairs and lobbying in the EU, the EU environment and climate change policy, the EU health policy and the EU research and innovation policy. Previously Pavlina has worked in the European Commission, DG Research, unit A5: Impact assessment of the EU Framework Programmes for Research and Technological development, and in a Brussels–based consultancy firm, Squaris Consultants Sprl, dealing with public affairs and EU projects.

Nick Sitter is a Professor of Political Economy at the BI Norwegian Business School and Professor of Public Policy at the Central European University. He holds a Ph.D. from the London School of Economics. Research interests include European integration, energy policy, party politics and political violence; books include *Europe's Nascent State: Public Policy in the European Union* (Gyldendal, 2006), *Understanding Public Management* (Sage, 2008), *A Liberal Actor in a Realist World: The EU Regulatory State and the Global Political Economy of Energy* (Oxford University Press, 2015) and *Terrorismens Historie* (Dreyer, 2015).

Marit Sjøvaag Marino is a political scientist whose main research interest is climate policy both in Norway and internationally. She holds a Ph.D. from the London School of Economics and Political Science. In the period 2008–2011 she was responsible for the annual Status Report on the implementation of climate policies in Norway at the Norwegian Business School's Center for Climate Strategy. She currently works on energy and climate issues both in Norway and in an international context for the Norwegian NGO Bellona.

John D. Sterman is the Jay W. Forrester Professor of Management and the Director of the System Dynamics Group at the MIT Sloan School of Management. His research includes systems thinking and organizational learning, computer simulation of corporate strategy and public policy issues, and environmental sustainability. He is the author of many scholarly and popular articles on the challenges and opportunities facing society and organizations today, including the book, *Modeling for Organizational Learning*, and the award-winning textbook, *Business Dynamics*.

Per Espen Stoknes is a psychologist with a Ph.D. in economics, who co-chairs with Jorgen Randers the Center for Climate Strategy at the Norwegian Business School. He also spearheads the Business school's Master of Management programs and research on *Green Growth*. A serial entrepreneur, including co-founder of the clean-tech company GasPlas, he's also written several books, the latest titled *What We Think About When We Try Not To Think About Global Warming* (2015).

Kristin Thorsrud Teien is senior adviser at the Norwegian Ministry of Climate and Environment, with responsibility for biodiversity policy and development of the Nature Index. She was previously affiliated with WWF Norway, leading the work on implementing the Living Planet Index for Norway.

Mathis Wackernagel is co-creator of the Ecological Footprint and President of Global Footprint Network. He completed a Ph.D. in community and regional planning with Professor William Rees at the University of British Columbia, where his doctoral dissertation developed the Ecological Footprint concept. Mathis has authored and contributed to more than 50 peer-reviewed papers, numerous articles and reports and various books on sustainability that focus on embracing resource limits and developing metrics for sustainability, including *Our Ecological Footprint: Reducing Human Impact on the Earth; Sharing Nature's Interest*.